More Praise for *Stiff*

"[Mary] Roach is authoritative, endlessly curious and drolly funny. Her research is scrupulous and winningly presented."
—Adam Woog, *Seattle Times*

"Mary Roach is one of an endangered species: a science writer with a sense of humor. She is able to make macabre funny without looting death of its dignity."
—Brian Richard Boylan, *Denver Post*

"Roach's conversational tone and her gallows humor bring her subjects to life." —Alex Abramovich, *People*

" 'Uproariously funny' doesn't seem a likely description for a book on cadavers. However, Roach . . . has done the nearly impossible and written a book as informative and respectful as it is irreverent and witty." —*Publishers Weekly*, starred review

"Every detail is riveting. It is impossible to tear one's eyes away from Roach's description." —Henry Kisor, *Chicago Sun-Times*

"[Roach] has written a curiously funny, touching, respectful study. . . . [She] bravely goes where we wouldn't want to go."
—Nancy Summers, *Tampa Tribune*

"Roach writes in an insouciant style and displays her métier in tangents about bizarre incidents in pathological history. Death may have the last laugh, but, in the meantime, Roach finds merriment in the macabre." —*Booklist*

"Droll, dark, and quite wise, *Stiff* makes being dead funny and fascinating and weirdly appealing."
—Susan Orlean, best-selling author of *The Orchid Thief*

"As fascinating as it is funny, as sensitive as it is probing, Mary Roach's *Stiff* is above all an important account of how we treat the dead—literally. The research is admirable, the anecdotes carefully chosen, and the prose lively."
—Caleb Carr, author of *The Alienist*

"Mary Roach is the funniest science writer in the country. . . . [*Stiff*] manages, somehow, to find humor in cadavers without robbing them of their dignity."
—Burkhard Bilger, author of *Noodling for Flatheads*

ALSO BY MARY ROACH

Fuzz: When Nature Breaks the Law

Grunt: The Curious Science of Humans at War

Gulp: Adventures on the Alimentary Canal

Packing for Mars: The Curious Science of Life in the Void

Bonk: The Curious Coupling of Science and Sex

Spook: Science Tackles the Afterlife

STIFF

The Curious Lives of Human Cadavers

Mary Roach

W. W. NORTON & COMPANY

Independent Publishers Since 1923

For wonderful Ed

In several instances, names have been changed to protect privacy.
Photo credits: title page: Hulton Deutch Collection/Corbis; p. 17: Getty Images/Juliette
Lasserre; p. 35: Getty Images/Robin Lynn Gibson; p. 59: Photofest; p. 85: Photofest;
p. 111: Getty Images/Stephen Swintek; p. 129: Hulton Deutsch Collection/Corbis; p. 155:
Geoffrey Clements/Corbis; p. 165: Getty Images/Tracy Montana/Photolink; p. 197: Bettmann/
Corbis; p. 219: Getty Images/John A. Rizzo; p. 249: Minnesota Historical
Society/Corbis; p. 279: T.R. Tharp/Corbis

For information about permission to reproduce
selections from this book, write to Permissions,

W. W. Norton & Company, Inc., 500 Fifth Avenue,
New York, NY 10110

Manufacturing by LSC Communications
Book design by JAM Design
Production manager: Amanda Morrison

Library of Congress Cataloging-in-Publication Data
Roach, Mary.
Stiff : the curious lives of human cadavers / by Mary Roach.
p. cm.
Includes bibliographical references.
ISBN 0-393-05093-9
1. Human experimentation in medicine. 2. Dead. 3. Human dissection. I. Title.
R853.H8 R635 2003
611—dc21   2002152908

ISBN 978-0-393-88172-1 pbk.

W. W. Norton & Company, Inc., 500 Fifth Avenue, New York, N.Y. 10110
www.wwnorton.com

W. W. Norton & Company Ltd.
15 Carlisle Street, London W1D 3BS

1 2 3 4 5 6 7 8 9 0

Contents

———

Introduction

The way I see it, being dead is not terribly far off from being on a cruise ship. Most of your time is spent lying on your back. The brain has shut down. The flesh begins to soften. Nothing much new happens, and nothing is expected of you.

If I were to take a cruise, I would prefer that it be one of those research cruises, where the passengers, while still spending much of the day lying on their backs with blank minds, also get to help out with a scientist's research project. These cruises take their passengers to unknown, unimagined places. They give them the chance to do things they would not otherwise get to do.

I guess I feel the same way about being a corpse. Why lie around on your back when you can do something interesting and new, something *useful*? For every surgical procedure developed, from heart transplants to bunion surgery, cadavers have been there alongside the surgeons, making history in their own quiet, sundered way. For two thousand years, cadavers—some willingly, some unwittingly—have been involved in science's boldest strides and weirdest undertakings. Cadavers were around to help test France's first guillotine, the "humane" alternative to hanging. They were there at the labs of Lenin's embalmers, helping test the latest techniques. They've been there (on paper) at Congressional hearings, helping make the case for mandatory seat belts. They've ridden the Space Shuttle (okay, pieces of them), helped a graduate student in Tennessee debunk spontaneous human combustion, been crucified in a Parisian laboratory to test the authenticity of the Shroud of Turin.

In exchange for their experiences, these cadavers agree to a sizable amount of gore. They are dismembered, cut open, rearranged. But here's the thing: They don't *endure* anything. Cadavers are our superheroes: They brave fire without flinching, withstand falls from tall buildings and head-on car crashes into walls. You can fire a gun at them or run a speedboat over their legs, and it will not faze them. Their heads can be removed with no deleterious effect. They can be in six places at once. I take the Superman point of view: What a shame to waste these powers, to not use them for the betterment of humankind.

This is a book about notable achievements made while dead. There are people long forgotten for their contributions while alive, but immortalized in the pages of books and journals. On my wall is a calendar from the Mütter Museum at the College of Physicians of Philadelphia. The photograph for October is of a piece of human skin, marked up with arrows and tears; it was used by surgeons to figure out whether an incision would be less likely to tear if it ran lengthwise or crosswise. To me, ending up an exhibit in the Mütter Museum or a skeleton in a medical school classroom is like donating money for a park bench after you're gone: a nice thing to do, a little hit of immortality. This is a book about the sometimes odd, often shocking, always compelling things cadavers have done.

Not that there's anything wrong with just lying around on your back. In its way, rotting is interesting too, as we will see. It's just that there are other ways to spend your time as a cadaver. Get involved with science. Be an art exhibit. Become part of a tree. Some options for you to think about.

Death. It doesn't have to be boring.

There are those who will disagree with me, who feel that to do anything other than bury or cremate the dead is disrespectful. That includes, I suspect, writing about them. Many people will find this book disrespectful. There is nothing amusing about being dead, they will say. Ah, but there is. Being dead is absurd. It's the silliest situation you'll find yourself in. Your limbs are floppy and uncooperative. Your mouth hangs open. Being dead is unsightly and stinky and embarrassing, and there's not a damn thing to be done about it.

This book is not about death as in dying. Death, as in dying, is sad and profound. There is nothing funny about losing someone you love, or about being the person about to be lost. This book is about the already dead, the anonymous, behind-the-scenes dead. The cadavers I have seen were not depressing or heart-wrenching or repulsive. They seemed sweet and well-intentioned, sometimes sad, occasionally amusing. Some were beautiful, some monsters. Some wore sweatpants and some were naked, some in pieces, others whole.

All were strangers to me. I would not want to watch an experiment, no matter how interesting or important, that involved the remains of someone I knew and loved. (There are a few who do. Ronn Wade, who runs the anatomical gifts program at the University of Maryland at Baltimore, told me that some years back a woman whose husband had willed his body to the university asked if she could watch the dissection. Wade gently said no.) I feel this way not because what I would be watching is disrespectful, or wrong, but because I could not, emotionally, separate that cadaver from the person it recently was. One's own dead are more than cadavers, they are place holders for the living. They are a focus, a receptacle, for emotions that no longer have one. The dead of science are always strangers.*

Let me tell you about my first cadaver. I was thirty-six, and it was eighty-one. It was my mother's. I notice here that I used the possessive "my mother's," as if to say the cadaver that belonged to my mother, not the cadaver that *was* my mother. My mom was never a cadaver; no person ever is. You are a person and then you cease to be a person, and a cadaver takes your place. My mother was gone. The cadaver was her hull. Or that was how it seemed to me.

It was a warm September morning. The funeral home had told me and my brother Rip to show up there about an hour before the church service. We thought there were papers to fill out. The mortician ushered us into a large, dim, hushed room with heavy drapes and too much air-conditioning. There was a coffin at one end, but this seemed normal enough, for a mortuary. My brother and I stood there awkwardly. The mortician cleared his throat and looked toward the coffin. I suppose we should have recognized it, as we'd picked it out and paid for it the day before, but we didn't. Finally the man walked over and gestured at it, bowing slightly, in the manner of a maître d' showing diners to their table. There, just beyond his open palm, was our mother's face. I wasn't expecting it. We hadn't requested a viewing, and the memorial service was closed-coffin. We got it anyway. They'd shampooed and waved her

---

* Or almost always. Every now and then, it will happen that an anatomy student recognizes a lab cadaver. "I've had it happen twice in a quarter of a century," says Hugh Patterson, an anatomy professor at the University of California, San Francisco, Medical School.

hair and made up her face. They'd done a great job, but I felt taken, as if we'd asked for the basic carwash and they'd gone ahead and detailed her. Hey, I wanted to say, we didn't order this. But of course I said nothing. Death makes us helplessly polite.

The mortician told us we had an hour with her, and quietly retreated. Rip looked at me. An *hour*? What do you do with a dead person for an hour? Mom had been sick for a long time; we'd done our grieving and crying and saying goodbye. It was like being served a slice of pie you didn't want to eat. We felt it would be rude to leave, after all the trouble they'd gone to. We walked up to the coffin for a closer look. I placed my palm on her forehead, partly as a gesture of tenderness, partly to see what a dead person felt like. Her skin was cold the way metal is cold, or glass.

A week ago at that time, Mom would have been reading the *Valley News* and doing the Jumble. As far as I know, she'd done the Jumble every morning for the past forty-five years. Sometimes in the hospital, I'd get up on the bed with her and we'd work on it together. She was bedridden, and it was one of the last things she could still do and enjoy. I looked at Rip. Should we all do the Jumble together one last time? Rip went out to the car to get the paper. We leaned on the coffin and read the clues aloud. That was when I cried. It was the small things that got to me that week: finding her bingo winnings when we cleaned out her dresser drawers, emptying the fourteen individually wrapped pieces of chicken from her freezer, each one labeled "chicken" in her careful penmanship. And the Jumble. Seeing her cadaver was strange, but it wasn't really sad. It wasn't *her*.

What I found hardest to get used to this past year was not the bodies I saw, but the reactions of people who asked me to tell them about my book. People want to be excited for you when they hear you are writing a book; they want to have something nice to say. A book about dead bodies is a conversational cur- veball. It's all well and good to write an article about corpses, but a full-size book plants a red flag on your character. *We knew Mary was quirky, but now we're wondering if she's, you know, okay.* I experienced a moment last summer at the checkout desk at the medical school library at the University of California, San Francisco, that sums up what it is like to write a book about cadavers. A young man was looking at the computer record of the books under my name: *The Principles and Prac- tice of Embalming, The Chemistry of Death, Gunshot Injuries.* He looked at the book I now wished to check out: *Proceedings of the Ninth Stapp Car Crash Conference.* He didn't say anything, but he didn't need to. It was all there in his glance. Often when I checked out a book I expected to be questioned. Why do you want this book? What are you up to? What kind of person are you?

They never asked, so I never told them. But I'll tell you now. I'm a curious person. Like all journalists, I'm a voyeur. I write about what I find fascinating. I used to write about travel. I traveled to escape the known and the ordinary. The longer I did this, the farther afield I had to go. By the time I found myself in Antarctica for the third time, I began to search closer at hand. I began to look for the foreign lands between the cracks. Science was one such land. Science involving the dead was particularly foreign and strange and, in its repellent way, enticing. The places I traveled to this past year were not as beautiful as Antarctica, but they were as strange and inter- esting and, I hope, as worthy of sharing.

STIFF

1

A HEAD IS A TERRIBLE
THING TO WASTE

Practicing surgery on the dead

The human head is of the same approximate size and weight as a roaster chicken. I have never before had occasion to make the comparison, for never before today have I seen a head in a roasting pan. But here are forty of them, one per pan, resting face-up on what looks to be a small pet-food bowl. The heads are for plastic surgeons, two per head, to practice on. I'm observing a facial anatomy and face-lift refresher course, sponsored by a southern university medical center and led by a half-dozen of America's most sought-after face-lifters.

The heads have been put in roasting pans—which are of the disposable aluminum variety—for the same reason chickens are put in roasting pans: to catch the drippings. Surgery, even surgery upon the dead, is a tidy, orderly affair. Forty folding utility tables have been draped in lavender plastic cloths, and a roasting pan is centered on each. Skin hooks and retractors are set out with the pleasing precision of restaurant cutlery. The whole thing has the look of a catered reception. I mention to the young woman whose job it was to set up the seminar this morning that the lavender gives the room a cheery sort of Easter-party feeling. Her name is Theresa. She replies that lavender was chosen because it's a soothing color.

It surprises me to hear that men and women who spend their

days pruning eyelids and vacuuming fat would require anything in the way of soothing, but severed heads can be upsetting even to professionals. Especially fresh ones ("fresh" here meaning unembalmed). The forty heads are from people who have died in the past few days and, as such, still look very much the way they looked while those people were alive. (Embalming hardens tissues, making the structures less pliable and the surgery experience less reflective of an actual operation.)

For the moment, you can't see the faces. They've been draped with white cloths, pending the arrival of the surgeons. When you first enter the room, you see only the tops of the heads, which are shaved down to stubble. You could be looking at rows of old men reclining in barber chairs with hot towels on their faces. The situation only starts to become dire when you make your way down the rows. Now you see stumps, and the stumps are not covered. They are bloody and rough. I was picturing something cleanly sliced, like the edge of a deli ham. I look at the heads, and then I look at the lavender tablecloths. Horrify me, soothe me, horrify me.

They are also very short, these stumps. If it were my job to cut the heads off bodies, I would leave the neck and cap the gore somehow. These heads appear to have been lopped off just below the chin, as though the cadaver had been wearing a turtleneck and the decapitator hadn't wished to damage the fabric. I find myself wondering whose handiwork this is.

"Theresa?" She is distributing dissection guides to the tables, humming quietly as she works.

"Mm?"

"Who cuts off the heads?"

Theresa answers that the heads are sawed off in the room across the hall, by a woman named Yvonne. I wonder out loud whether this particular aspect of Yvonne's job bothers her. Likewise Theresa. It was Theresa who brought the heads in and set them up on their little stands. I ask her about this.

"What I do is, I think of them as wax."

Theresa is practicing a time-honored coping method: objectification. For those who must deal with human corpses regularly, it is easier (and, I suppose, more accurate) to think of them as objects, not people. For most physicians, objectification is mastered their first year of medical school, in the gross anatomy lab, or "gross lab," as it is casually and somewhat aptly known. To help depersonalize the human form that students will be expected to sink knives into and eviscerate, anatomy lab personnel often swathe the cadavers in gauze and encourage students to unwrap as they go, part by part.

The problem with cadavers is that they look so much like people. It's the reason most of us prefer a pork chop to a slice of whole suckling pig. Dissection and surgical instruction, like meat-eating, require a carefully maintained set of illusions and denial. Physicians and anatomy students must learn to think of cadavers as wholly unrelated to the people they once were. "Dissection," writes historian Ruth Richardson in *Death, Dissection, and the Destitute,* "requires in its practitioners the effective suspension or suppression of many normal physical and emotional responses to the wilful mutilation of the body of another human being."

Heads—or more to the point, faces—are especially unsettling. At the University of California, San Francisco, in whose medical school anatomy lab I would soon spend an afternoon, the head and hands are often left wrapped until their dissection comes up on the syllabus. "So it's not so intense," one student would later tell me. "Because that's what you see of a person."

The surgeons are beginning to gather in the hallway outside the lab, filling out paperwork and chatting volubly. I go out to watch them. Or to not watch the heads, I'm not sure which. No one pays much attention to me, except for a small, dark-haired woman, who stands off to the side, staring at me. She doesn't look as if she wants to be my friend. I decide to think of her as wax. I talk with the surgeons, most of whom seem to think I'm part of the setup staff. A man with a shrubbery of white chest hair in the V-neck of his surgical scrubs says to me: "Were y'in there injectin' 'em with water?" A Texas accent makes taffy of his syllables. "Plumpin' 'em up?" Many of today's heads have been around a few days and have, like any refrigerated meat, begun to dry out. Injections of saline, he explains, are used to freshen them.

Abruptly, the hard-eyed wax woman is at my side, demanding to know who I am. I explain that the surgeon in charge of the symposium invited me to observe. This is not an entirely truthful rendering of the events. An entirely truthful rendering of the events would employ words such as "wheedle," "plead," and "attempted bribe."

"Does publications know you're here? If you're not cleared through the publications office, you'll have to leave." She strides into her office and dials the phone, staring at me while she talks, like security guards in bad action movies just before Steven Seagal clubs them on the head from behind.

One of the seminar organizers joins me. "Is Yvonne giving you a hard time?"

Yvonne! My nemesis is none other than the cadaver beheader. As it turns out, she is also the lab manager, the person responsible when things go wrong, such as writers fainting and/or getting sick to their stomach and then going home and writing books that refer to anatomy lab managers as beheaders. Yvonne is off the phone now. She has come over to outline her misgivings. The seminar organizer reassures her. My end of the conversation takes place entirely in my head and consists of a single repeated line. *You cut off heads. You cut off heads. You cut off heads.*

Meanwhile, I've missed the unveiling of the faces. The surgeons are already at work, leaning kiss-close over their specimens and glancing up at video monitors mounted above each work station. On the screen are the hands of an unseen narrator, demonstrating the procedures on a head of his own. The shot is an extreme close-up, making it impossible to tell, without already knowing, what kind of flesh it is. It could be Julia Child skinning poultry before a studio audience.

The seminar begins with a review of facial anatomy. "Elevate the skin in a subcutaneous plane from lateral to medial," intones the narrator. Obligingly, the surgeons sink scalpels into faces. The flesh gives no resistance and yields no blood.

"Isolate the brow as a skin island." The narrator speaks slowly, in a flat tone. I'm sure the idea is to sound neither excited and delighted at the prospect of isolating skin islands, nor overly dismayed. The net effect is that he sounds chemically sedated, which seems to me like a good idea.

I walk up and down the rows. The heads look like rubber Halloween masks. They also look like human heads, but my brain has no precedent for human heads on tables or in roasting pans or anywhere other than on top of human bodies, and so I think it has chosen to interpret the sight in a more comforting manner. *Here we are at the rubber mask factory. Look at the nice men and women working on the masks.* I used to have a Halloween mask of an old toothless man whose lips fell in upon his gums. There are several of him here. There is a Hunchback of Notre Dame, bat-nosed and with lower teeth exposed, and a Ross Perot.

The surgeons don't seem queasy or repulsed, though Theresa told me later that one of them had to leave the room. "They hate it," she says. "It" meaning working with heads. I sense from them only a mild discomfort with their task. As I stop at their tables to watch, they turn to me with a vaguely irritated, embarrassed look. You've seen that look if you make a habit of entering bathrooms without knocking. The look says, Please go away.

Though the surgeons clearly do not relish dissecting dead people's heads, they just as clearly value the opportunity to practice and explore on someone who isn't going to wake up and look in the mirror anytime soon. "You have a structure you keep seeing [during surgeries], and you're not sure what it is, and you're afraid to cut it," says one surgeon. "I came here with four questions." If he leaves today with answers, it will have been worth the $500. The surgeon picks his head up and sets it back down, adjusting its position like a seamstress pausing to shift the cloth she is working on. He points out that the heads aren't cut off out of ghoulishness. They are cut off so that someone else can make use of the other pieces: arms, legs, organs. In the world of donated cadavers, nothing is wasted. Before their face-lifts, today's heads got nose jobs in the Monday rhinoplasty lab.

It's the nose jobs that I trip over. Kindly, dying southerners willed their bodies for the betterment of science, only to end up as practice runs for nose jobs? Does it make it okay that the kindly southerners, being dead kindly southerners, have no way of knowing that this is going on? Or does the deceit compound the crime? I spoke about this later with Art Dalley, the director of the Medical Anatomy Program at Vanderbilt University in Nashville and an expert in the history of anatomical gift-giving. "I think there's a surprising number of donors who really don't care what happens to them," Dalley told me. "To them it's just a practical means of disposing of a body, a practical means that fortunately has a ring of altruism."

Though it's harder to justify the use of a cadaver for practicing nose jobs than it is for practicing coronary bypasses, it is justifiable nonetheless. Cosmetic surgery exists, for better or for worse, and it's important, for the sake of those who undergo it, that the surgeons who do it are able to do it well. Though perhaps there ought to be a box for people to check, or not check, on their body donor form: *Okay to use me for cosmetic purposes.*★

I sit down at Station 13, with a Canadian surgeon named Marilena Marignani. Marilena is dark-haired, with large eyes and strong cheekbones. Her head (the one on the table) is gaunt, with a similarly strong set to the bones. It's an odd way for the two women's lives to intersect; the head doesn't need a face-lift, and Marilena doesn't usually do them. Her practice is primarily reconstructive plastic surgery. She has done only two face-lifts before and wants to hone her skills before undertaking a procedure on a friend. She wears a mask over her nose and mouth, which is surprising, because a severed head is in no danger of infection. I ask whether this is more for her own protection, a sort of psychological barrier.

Marilena replies that she doesn't have a problem with heads. "For me, hands are hard." She looks up from what she's doing. "Because you're holding this disconnected hand, and it's holding you back." Cadavers occasionally effect a sort of accidental humanness that catches the medical professional off guard. I once spoke to an anatomy student who described a moment in the lab when she realized the cadaver's arm was around her waist. It becomes difficult, under circumstances such as these, to retain one's clinical remove.

I watch Marilena gingerly probing the woman's exposed tissue. What she is doing, basically, is getting her bearings: learning—in a detailed, hands-on manner—what's what and what's where in the complicated layering of skin, fat, muscle, and fascia that makes up the human cheek. While early face-lifts merely pulled the skin up and stitched it, tightened, into

---

★ I'm a believer in organ and tissue (bone, cartilage, skin) donation, but was startled to learn that donated skin that isn't used for, say, grafting onto burn victims may be processed and used cosmetically to plump up wrinkles and aggrandize penises. While I have no preconceived notions of the hereafter, I stand firm in my conviction that it should not take the form of someone else's underpants.

place, the modern face-lift lifts four individual anatomical layers. This means all of these layers must be identified, surgically separated from their neighbors, individually repositioned, and sewn into place—all the while taking care not to sever vital facial nerves. With more and more cosmetic procedures being done endoscopically—by introducing tiny instruments through a series of minimally invasive incisions—knowing one's way around the anatomy is even more critical. "With the older techniques, they peeled everything down and they could see it all in front of them," says Ronn Wade, director of the Anatomical Services Division of the University of Maryland School of Medicine. "Now when you go in with a camera and you're right on top of something, it's harder to keep yourself oriented."

Marilena's instruments are poking around the edges of a glistening yolk-colored blob. The blob is known among plastic surgeons as the malar fat pad. "Malar" means relating to the cheek. The malar fat pad is the cushion of youthful padding that sits high on your cheekbone, the thing grandmothers pinch. Over the years, gravity coaxes the fat from its roost, and it commences a downward slide, piling up at the first anatomical roadblock it reaches: the nasolabial folds (the anatomical parentheses that run from the edges of a middle-aged nose down to the corners of the mouth). The result is that the cheeks start to look bony and sunken, and bulgy parentheses of fat reinforce the nasolabial lines. During face-lifts, surgeons put the malar fat pad back up where it started out.

"This is great," says Marilena. "Beautiful. Just like real, but no bleeding. You can really see what you're doing."

Though surgeons in all disciplines benefit from the chance to try out new techniques and new equipment on cadaveric specimens, fresh parts for surgical practice are hard to come by. When I telephoned Ronn Wade in his office in Baltimore, he explained that the way most willed body programs are set up, anatomy labs have first priority when a cadaver comes in. And even when there's a surplus, there may be no infrastructure in place to get the bodies from the anatomy department of the medical school over to the hospitals where the surgeons are—and no place at the hospital for a surgical practice lab. At Marilena's hospital, surgeons typically get body parts only when there's been an amputation. Given the frequency of human head amputations, an opportunity like today's would be virtually nonexistent outside of a seminar.

Wade has been working to change the system. He is of the opinion—and it's hard to disagree with him—that live surgery is the worst place for a surgeon to be practicing a new skill. So he got together with the heads—sorry, *chiefs*—of surgery at Baltimore's hospitals and worked out a system. "When a group of surgeons want to get together and try out, say, some new endoscopic technique, they call me and I set it up." Wade charges a nominal fee for the use of the lab, plus a small per-cadaver fee. Two-thirds of the bodies Wade takes in now are being used for surgical practice.

I was surprised to learn that even when surgeons are in residencies, they aren't typically given an opportunity to practice operations on donated cadavers. Students learn surgery the way they have always learned: by watching experienced surgeons at work. At teaching hospitals affiliated with medical schools, patients who undergo surgery typically have an audience of interns. After watching an operation a few times, the intern is invited to step in and try his or her hand, first on simple maneuvers such as closures and retractions, and gradually at more complicated steps. "It's basically on-the-job training," says Wade. "It's an apprenticeship."

It has been this way since the early days of surgery, the teaching of the craft taking place largely in the operating room. Only in the past century, however, has the patient routinely stood to gain from the experience. Nineteenth-century operating "theaters" had more to do with medical instruction than with saving patients' lives. If you could, you stayed out of them at all cost.

For one thing, you were being operated on without anesthesia. (The first operations under ether didn't take place until 1846.) Surgical patients in the late 1700s and early 1800s could feel every cut, stitch, and probing finger. They were often blindfolded—this may have been optional, not unlike the firing squad hood—and invariably bound to the operating table to keep them from writhing and flinching or, quite possibly, leaping from the table and fleeing into the street. (Perhaps owing to the presence of an audience, patients underwent surgery with most of their clothes on.)

The early surgeons weren't the hypereducated cowboy-saviors they are today. Surgery was a new field, with much to be learned and near-constant blunders. For centuries, surgeons had shared rank with barbers, doing little beyond amputations and tooth pullings, while physicians, with their potions and concoctions, treated everything else. (Interestingly, it was proctology that helped pave the way for surgery's acceptance as a respectable branch of medicine. In 1687, the king of France was surgically relieved of a painful and persistent anal fistula and was apparently quite grateful for, and vocal about, his relief.)

Nepotism, rather than skill, secured a post at early-nineteenth-century teaching hospitals. The December 20, 1828, issue of The Lancet contains excerpts from one of the earliest surgical malpractice trials, which centered on the incompetency of one Bransby Cooper, nephew of the famed anatomist Sir Astley Cooper. Before an audience of some two hundred colleagues, students, and onlookers, the young Cooper proved beyond question that his presence in the operating theater owed everything to his uncle and nothing to his talents. The operation was a simple bladder stone removal (lithotomy) at London's Guy's Hospital; the patient, Stephen Pollard, was a hardy working-class man. While lithotomies were normally completed in a matter of minutes, Pollard was on the table for an hour, with his knees at his neck and his hands bound to his feet while the clueless medic tried in vain to locate the stone. "A blunt gorget was also introduced, and the scoop, and several pair of forceps," recalled one witness. Another described the "horrible squash, squash of the forceps in the perineum." When a succession of tools failed to produce the stone, Cooper "introduced his finger with some force. . . ." It was around this point that Pollard's endurance* ran dry. "Oh! Let it go!" he is quoted as saying. "Pray let it keep in!" Cooper persisted, cursing the man's deep perineum (in fact, an autopsy showed it to be a quite normally proportioned perineum). After digging with his finger for some ungodly amount of time, he got up from his seat and "measured fingers with those of other gentlemen, to see if any of them had a longer finger." Eventually he went back to his toolkit and, with forceps, conquered the recalcitrant rock—a relatively small one, "not larger than a common Windsor bean"—brandishing it above his head like an Academy Award winner. The quivering, exhausted mass

---

* The human being of centuries past was clearly in another league, insofar as pain endurance went. The farther back you go, it seems, the more we could take. In medieval England, the patient wasn't even tied down, but sat obligingly upon a cushion at the foot of the doctor's chair, presenting his ailing part for treatment. In an illustration in *The Medieval Surgery*, we find a well-coiffed man about to receive treatment for a troublesome facial fistula. The patient is shown calmly, almost fondly, lifting his afflicted face toward the surgeon. Meanwhile, the caption is going: "The patient is instructed to avert his eyes and . . . the roots of the fistula are then seared by taking an iron or bronze tube through which is passed a red hot iron." The caption writer adds, "The doctor would appear to be left-handed in this particular picture," as if perhaps trying to distract the reader from the horrors just read, a palliative technique fully as effective as asking a man with a red-hot poker closing in on his face to "avert his eyes."

that was Stephen Pollard was wheeled to a bed, where he died of infection and God knows what else twenty-nine hours later.

Bad enough that some ham-handed fop in a waistcoat and bowtie was up to his wrists in your urinary tract, but on top of that you had an audience—not just the young punters from the medical school but, judging from a description of another lithotomy at Guy's Hospital in an 1829 *Lancet,* half the city: "Surgeons and surgeons' friends, . . . French visitors, and interlopers filled the space around the table. . . . There was soon a general outcry throughout the gallery and upper rows—'hat's off,' 'down heads,' . . . was loudly vociferated from different parts of the theatre."

The cabaret atmosphere of early medical instruction began centuries before, in the standing-room-only dissecting halls of the renowned Italian medical academies of Padua and Bologna. According to C. D. O'Malley's biography of the great Renaissance anatomist Andreas Vesalius, one enthusiastic spectator at a crowded Vesalius dissection, bent on a better view, leaned too far out and tumbled from his bench to the dissecting platform below. "Because of his accidental fall . . . , the unfortunate Master Carlo is unable to attend and is not very well," read the note proffered at the next lecture. Master Carlo, one can be sure, did not seek treatment at the place he went for lectures.

Without exception, the only people who checked themselves in at teaching hospitals were those too poor to pay for private surgery. In return for an operation that was as likely to kill them as make them better—bladder stone removal had a mortality rate of 50 percent—the poor basically donated themselves as living practice material. Not only were the surgeons unskilled, but many of the operations being done were purely experimental—no one really expected them to help. Wrote historian Ruth Richardson in *Death, Dissection, and the Destitute,* "The benefit [to the patient] was often incidental to the experiment."

With the advent of anesthesia, patients were at least unconscious while the young resident tried his hand at a new procedure. But they probably didn't give their permission for a trainee to take the helm. In the heady days before consent forms and drop-of-a-hat lawsuits, patients didn't realize what they might be in for if they underwent surgery at a teaching hospital, and doctors took advantage of this fact. While a patient was under, a surgeon might invite a student to practice an appendectomy. Never mind that the patient didn't have appendicitis. One of the more common transgressions was the gratuitous pelvic exam. A budding M.D.'s first Pap smear—the subject of significant anxiety and dread—was often administered to an unconscious female surgical patient. (Nowadays, enlightened medical schools will hire a "pelvic educator," a sort of professional vagina who allows the students to practice on her and offers personalized feedback and is, in my book anyway, a nominee for sainthood.)

Gratuitous medical procedures happen far less than they used to, owing to the public's growing awareness. "Patients are savvier these days, and the climate has changed a great deal," Hugh Patterson, who runs the willed body program at the University of California, San Francisco, Medical School, told me. "Even at a teaching hospital, patients request that residents not do the surgery. They want to be assured the attending does the procedure. It makes training very difficult."

Patterson would like to see specialized cadaver anatomy labs added to third- and fourth-year programs—instead of teaching anatomy only in the first year, "as one big bolus." Already, he and his colleagues have added a focused dissection, similar to the facial anatomy lab I'm observing today, to the curricula of surgical subspecialties. They've also set up a series of sessions at the medical school morgue to teach emergency room procedures to third-year students. Before a cadaver is embalmed and delivered to the anatomy lab, it may pass an afternoon getting tracheal intubations and catheterizations. (Some schools use anesthetized dogs for this purpose.) Given the urgency and difficulty of certain ER procedures, it makes good sense to practice them first on the dead. In the past, this has been done in a less formal manner, on freshly dead hospital patients, without consent—a practice whose propriety is intermittently debated in hushed meetings of the American Medical Association. They should probably just ask for permission: According to one *New England Journal of Medicine* study on the subject, 73 percent of parents of newly dead children, when asked, gave consent to use their child's body for teaching intubation skills.

I ask Marilena if she plans to donate her remains. I have always assumed that a sense of reciprocity prompts doctors to donate—repayment for the generosity of the people they dissected in medical school. Marilena, for one, isn't going to. She cites a lack of respect. It surprises me to hear her say this. As far as I can tell, the heads are being treated with respect. I hear no joking or laughter or callous comments. If there can be a respectful way to "deglove" a face, if loosening the skin of someone's forehead and flipping it back over his or her eyes can be a respectful act, then I think these people are managing it. It's strictly business.

It turns out that what Marilena objected to was a couple of the surgeons' taking photographs of their cadaver heads. When you take a photograph of a patient for a medical journal, she points out, you have the patient sign a release. The dead can't refuse to sign releases, but that doesn't mean they wouldn't want to. This is why cadavers in photographs in pathology and forensics journals have black bars over their eyes, like women on the Dos and Don'ts pages of *Glamour*. You have to assume that people don't want to be photographed dead and dismembered, any more than they want to be photographed naked in the shower or asleep on a plane with their mouth hanging open.

Most doctors aren't worried about a lack of respect from other doctors. Most of the ones I've spoken to would worry, if anything, about a lack of respect from students in the first-year gross anatomy lab—my next stop.

The seminar is nearly over. The video monitors are blank and the surgeons are cleaning up and filing out into the hallway. Marilena replaces the white cloth on her cadaver's face; about half the surgeons do this. She is conscientiously respectful. When I asked her why the eyes of the dead woman had no pupils, she did not answer, but reached up and closed the eyelids. As she slides back her chair, she looks down at the benapkined form and says, "May she rest in peace." I hear it as "pieces," but that's just me.

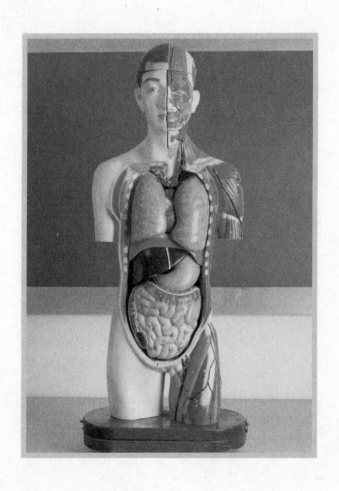

2

CRIMES OF ANATOMY

Body snatching and other sordid tales
from the dawn of human dissection

Enough years have passed since the use of Pachelbel's Canon
in a fabric softener commercial that the music again sounds
pure and sweetly sad to me. It's a good choice for a memorial
service, a classic and effective choice, for the men and women
gathered (here today) have fallen silent and somber with the
music's start.

Noticeably absent amid the flowers and candles is the casket displaying the deceased. This would have been logistically
challenging, as all twenty-some corpses have been reduced to
neatly sawed segments—hemisections of pelvis and bisected
heads, the secret turnings of their sinus cavities revealed like
Ant Farm tunnels. This is a memorial service for the unnamed
cadavers of the University of California, San Francisco, Medical School Class of 2004 gross anatomy lab. An open-casket
ceremony would not have been especially horrifying for the
guests here today, for they have not only seen the deceased in
their many and various pieces, but have handled them and are
in fact the reason they have been dismembered. They are the
anatomy lab students.

This is no token ceremony. It is a sincere and voluntarily
attended event, lasting nearly three hours and featuring thir-

teen student tributes, including an a capella rendition of Green Day's "Time of Your Life," the reading of an uncharacteristically downbeat Beatrix Potter tale about a dying badger, and a folk ballad about a woman named Daisy who is reincarnated as a medical student whose gross anatomy cadaver turns out to be himself in a former life, i.e., Daisy. One young woman's tribute describes unwrapping her cadaver's hands and being brought up short by the realization that the nails were painted pink. "The pictures in the anatomy atlas did not show nail polish," she wrote. "Did you choose the color? . . . Did you think that I would see it? . . . I wanted to tell you about the inside of your hands. . . . I want you to know you are always there when I see patients. When I palpate an abdomen, yours are the organs I imagine. When I listen to a heart, I recall holding your heart." It is one of the most touching pieces of writing I've ever heard. Others must feel the same; there isn't an anhydrous lacrimal gland in the house.

Medical schools have gone out of their way in the past decade to foster a respectful attitude toward gross anatomy lab cadavers. UCSF is one of many medical schools that hold memorial services for willed bodies. Some also invite the cadavers' families to attend. At UCSF, gross anatomy students must attend a precourse workshop hosted by students from the prior year, who talk about what it was like to work with the dead and how it made them feel. The respect and gratitude message is liberally imparted. From what I've heard, it would be quite difficult, in good conscience, to attend one of these workshops and then proceed to stick a cigarette in your cadaver's mouth or jump rope with his intestines.

Hugh Patterson, anatomy professor and director of the university's willed body program, invited me to spend an afternoon at the gross anatomy lab, and I can tell you here and now that either the students were exceptionally well rehearsed for my visit or the program is working. With no prompting on my part, the students spoke of gratitude and preserving dignity, of having grown attached to their cadavers, of feeling bad about what they had to do to them. "I remember one of my teammates was just hacking him apart, digging something out," one girl told me, "and I realized I was patting his arm, going, 'It's okay, it's okay.'" I asked a student named Matthew whether he would miss his cadaver when the course ended, and he replied that it was actually sad when "just part of him left." (Halfway through the course, the legs are removed and incinerated to reduce the students' exposure to the chemical preservatives.)

Many of the students gave their cadavers names. "Not like Beef Jerky. Real names," said one student. He introduced me to Ben the cadaver, who, despite having by then been reduced to a head, lungs, and arms, retained an air of purpose and dignity. When a student moved Ben's arm, it was picked up, not grabbed, and set down gently, as if Ben were merely sleeping. Matthew went so far as to write to the willed body program office asking for biographical information about his cadaver. "I wanted to personalize it," he told me.

No one made jokes the afternoon I was there, or anyway not at the corpses' expense. One woman confessed that her group had passed comment on the "extremely large genitalia" of their cadaver. (What she perhaps didn't realize is that the embalming fluid pumped into the veins expands the body's erectile tissues, with the result that male anatomy lab cadavers may be markedly better endowed in death than they were in life.) Even then, reverence, not mockery, colored the remark.

As one former anatomy instructor said to me, "No one's taking heads home in buckets anymore."

To understand the cautious respect for the dead that pervades the modern anatomy lab, it helps to understand the extreme lack of it that pervades the field's history. Few sciences are as rooted in shame, infamy, and bad PR as human anatomy.

The troubles began in Alexandrian Egypt, circa 300 B.C. King Ptolemy I was the first leader to deem it a-okay for medical types to cut open the dead for the purpose of figuring out how bodies work. In part this had to do with Egypt's long tradition of mummification. Bodies are cut open and organs removed during the mummification process, so these were things the government and the populace were comfortable with. It also had to do with Ptolemy's extracurricular fascination with dissection. Not only did the king issue a royal decree encouraging physicians to dissect executed criminals, but, come the day, he was over at the anatomy room with his knives and smock, slitting and probing alongside the pros.

Trouble's name was Herophilus. Dubbed the Father of Anatomy, he was the first physician to dissect human bodies. While Herophilus was indeed a dedicated and tireless man of science, he seems to have lost his bearings somewhere along the way. Enthusiasm got the better of compassion and common sense, and the man took to *dissecting live criminals*. According to one of his accusers, Tertullian, Herophilus vivisected six hundred prisoners. To be fair, no eyewitness account or papyrus diary entries survive, and one wonders whether professional jealousy played a role. After all, no one was calling Tertullian the Father of Anatomy.

The tradition of using executed criminals for dissections persisted and hit its stride in eighteenth- and nineteenth-century Britain, when private anatomy schools for medical students began to flourish in the cities of England and Scotland. While the number of schools grew, the number of cadavers stayed roughly the same, and the anatomists faced a chronic shortage of material. Back then no one donated his body to science. The churchgoing masses believed in a literal, corporal rising from the grave, and dissection was thought of as pretty much spoiling your chances of resurrection: Who's going to open the gates of heaven to some slob with his entrails all hanging out and dripping on the carpeting? From the sixteenth century up until the passage of the Anatomy Act, in 1836, the only cadavers legally available for dissection in Britain were those of executed murderers.

For this reason, anatomists came to occupy the same terrain, in the public's mind, as executioners. Worse, even, for dissection was thought of, literally, as a punishment worse than death. Indeed, that—not the support and assistance of anatomists—was the authorities' main intent in making the bodies available for dissection. With so many relatively minor offenses punishable by death, the legal bodies felt the need to tack on added horrors as deterrents against weightier crimes. If you stole a pig, you were hung. If you killed a man, you were hanged *and then dissected*. (In the freshly minted United States of America, the punishable-by-dissection category was extended to include duelists, the death sentence clearly not posing much of a deterrent to the type of fellow who agrees to settle his differences by the dueling pistol.)

Double sentencing wasn't a new idea, but rather the latest variation on the theme. Before that, a murderer might be hanged and then drawn and quartered, wherein horses were tied to his limbs and spurred off in four directions, the resultant "quarters" being impaled on spikes and publicly displayed, as a colorful reminder to the citizenry of the ill-advisedness of crime. Dissection as a sentencing option for murderers was mandated, in 1752 Britain, as an alternative to postmortem gibbeting. Gibbeting—though it hits the ear like a word for happy playground chatter or perhaps, at worst, the cleaning of small game birds—is in fact a ghastly verb. To gibbet is to dip a corpse in tar and suspend it in a flat iron cage (the gibbet) in plain view of townsfolk while it rots and gets pecked apart by crows. A stroll through the square must have been a whole different plate of tamales back then.

In attempting to cope with the shortage of cadavers legally available for dissection, instructors at British and early American anatomy schools backed themselves into some unsavory corners. They became known as the kind of guys to whom you could take your son's amputated leg and sell it for beer money (37 1/2 cents, to be exact; it happened in Rochester, New York, in 1831). But students weren't going to pay tuition to learn arm and leg anatomy; the schools had to find whole cadavers or risk losing their students to the anatomy schools of Paris, where the unclaimed corpses of the poor who died at city hospitals could be used for dissection.

Extreme measures ensued. It was not unheard of for an anatomist to tote freshly deceased family members over to the dissecting chamber for a morning before dropping them off at the churchyard. Seventeenth-century surgeon-anatomist William Harvey, famous for discovering the human circulatory system, also deserves fame for being one of few medical men in history so dedicated to his calling that he could dissect his own father and sister.

Harvey did what he did because the alternatives—stealing the corpses of someone else's loved ones or giving up his research—were unacceptable to him. Modern-day medical students living under Taliban rule faced a similar dilemma, and, on occasion, have made similar choices. In a strict interpretation of Koranic edicts regarding the dignity of the human body, Taliban clerics forbid medical instructors to dissect cadavers or use skeletons—even those of non-Muslims, a practice other Islamic countries often allow—to teach anatomy. In January 2002, *New York Times* reporter Norimitsu Onishi interviewed a student at Kandahar Medical College who had made the anguishing decision to dig up the bones of his beloved grandmother and share them with his class-mates. Another student unearthed the remains of his former neighbor. "Yes, he was a good man," the student told Onishi. "Naturally I felt bad about taking his skeleton. . . . I thought that if twenty people could benefit from it, it would be good."

This sort of reasoned, pained sensitivity was rare in the hey-day of British anatomy schools. The far more common tactic was to sneak into a graveyard and dig up someone else's relative to study. The act became known as body snatching. It was a new crime, distinct from grave-robbing, which involved the pilfering of jewels and heirlooms buried in tombs and crypts of the well-to-do. Being caught in possession of a corpse's cufflinks was a crime, but being caught with the corpse itself carried no penalty. Before anatomy schools caught on, there were no laws on the books regarding the misappropriation of freshly dead humans. And why would there be? Up until that point, there had been little reason, short of necrophilia,* to undertake such a thing.

Some anatomy instructors mined the timeless affinity of university students for late-night pranks by encouraging their enrollees to raid graveyards and provide bodies for the class. At certain Scottish schools, in the 1700s, the arrangement was more formal: Tuition, writes Ruth Richardson, could be paid in corpses rather than cash.

---

* Which was also, up until 1965, not a crime in any U.S. state. When necrophil-ia's best-known modern-day practitioner, Sacramento mortuary worker Karen Greenlee, was caught absconding with a dead young man in 1979, she was fined for illegally driving a hearse but not for the act itself, as California had no statutes regarding sex with the dead. To date, only sixteen states have enacted necrophilia laws. The language used by each state reflects its particular character. While tac-iturn Minnesota refers to those who "carnally know a dead body," freewheeling Nevada spells it all out: "It is a felony to engage in cunnilingus, fellatio, or any intrusion of any part of a person's body, or any object manipulated or inserted by a person into the genital or anal openings of the body of another where the offender performs these acts on the dead body of a human being."

Other instructors took the dismal deed upon themselves. These were not low-life quacks. They were respectable members of their profession. Colonial physician Thomas Sewell, who went on to become the personal physician to three U.S. presidents and to found what is now George Washington University Medical School, was convicted in 1818 of digging up the corpse of a young Ipswich, Massachusetts, woman for the purposes of dissection.

And then there were the anatomists who paid someone else to go digging. By 1828, the demands of London's anatomy schools were such that ten full-time body snatchers and two hundred or so part-timers were kept busy throughout the dissecting "season." (Anatomy courses were held only between October and May, to avoid the stench and swiftness of summertime decomposition.) According to a House of Commons testimony from that year, one gang of six or seven resurrectionists, as they were often called, dug up 312 bodies. The pay worked out to about $1,000 a year—some five to ten times the earnings of the average unskilled laborer—with summers off.

The job was immoral, and ugly to be sure, but probably less unpleasant than it sounds. The anatomists wanted freshly dead bodies, so the smell wasn't really a problem. A body snatcher didn't have to dig up the entire grave, but rather just the top end of it. A crowbar would then be slipped under the coffin lid and wrenched upward, snapping off the top foot or so. The corpse was fished out with a rope around the neck or under the arms, and the dirt, which had been piled on a tarp, would be slid back in. The whole affair took less than an hour.

Many of the resurrectionists had held posts as gravediggers or assistants in anatomy labs, where they'd come into contact with the gangs and their doings. Drawn to the promise of higher pay and less confining hours, they abandoned legitimate posts to take up the shovel and sack. A few diary entries—transcribed from the anonymously written *Diary of a*

Resurrectionist—yield some insight into the sort of people we're talking about here:

> Tuesday 3rd (November 1811). Went to look out and brought the Shovils from Bartholow, . . . Butler and me came home intoxsicated.

> Tuesday 10th. Intoxsicated all day: at night went out & got 5 Bunhill Row. Jack all most buried.

> Friday 27th. . . . Went to Harps, got 1 large and took it to Jack's house, Jack, Bill, and Tom not with us, Geting drunk.

It is tempting to believe that the author's impersonal references to the corpses belie some sense of discomfort with his activities. He does not dwell upon their looks or muse about their sorry fate. He cannot bring himself to refer to the dead as anything other than a size or a gender. Only occasionally do the bodies merit a noun. (Most often "thing," as in "Thing bad," meaning "body decomposed.") But more likely it was simply the man's disinclination to sit down and write that accounts for the shorthand. Later entries show he couldn't even be bothered so spell out "canines," which appears as "Cns." (When a "thing bad," the "Cns" and other teeth were pulled and sold to dentists, for making dentures,[*] so as to keep the undertaking from being a complete loss.)

Body snatchers were common thugs; their motive, simple greed. But what of the anatomists? Who were these upstanding members of society who could commission the theft and semipublic mutilation of someone's dead grandmother? The best-known of the London surgeon-anatomists was Sir Astley

---

[*] How could people of the nineteenth century have allowed teeth from cadavers to be put into their mouths? The same way people from the twenty-first century can allow tissue from cadavers to be injected into their faces to fill wrinkles. They possibly didn't know and probably didn't care.

Cooper. In public, Cooper denounced the resurrectionists, yet
he not only sought out and retained their services, but encouraged those in his employ to take up the job. Thing bad.

Cooper was an outspoken defender of human dissection.
"He must mangle the living if he has not operated on the
dead" was his famous line. While his point is well taken and
the medical schools' plight was a difficult one, a little conscience would have served. Cooper was the type of man who
not only evinced no compunction about cutting up strangers' family members but happily sliced into his own former
patients. He kept in touch with the family doctors of those he
had operated on and, upon hearing of their passing, commissioned his resurrectionists to unearth them so that he might
have a look at how his handiwork had held up. He paid for
the retrieval of bodies of colleagues' patients known to have
interesting ailments or anatomical peculiarities. He was a man
in whom a healthy passion for biology seemed to have metas-
tasized into a sort of macabre eccentricity. In *Things for the Surgeon*, an account of body snatching by Hubert Cole, Sir Astley
is said to have painted the names of colleagues onto pieces of
bone and forced lab dogs to swallow them, so that when the
bone was extracted during the dog's dissection, the colleague's
name would appear in intaglio, the bone around the letters
having been eaten away by the dog's gastric acids. The items
were handed out as humorous gifts. Cole doesn't mention the
colleagues' reactions to the one-of-a-kind nameplates, but I
would hazard a guess that the men made an effort to enjoy
the joke and displayed the items prominently, at least when Sir
Astley came calling. For Sir Astley wasn't the sort of fellow
whose ill will you wanted to take with you to your grave. As
Sir Astley himself put it, "I can get anyone."

Like the resurrectionists, the anatomists were men who had
clearly been successful in objectifying, in their own minds at
least, the dead human body. Not only did they view dissection

and the study of anatomy as justification for unapproved disinterment, they saw no reason to treat the unearthed dead as entities worthy of respect. It didn't bother them that the corpses would arrive at their doors, to quote Ruth Richardson, "compressed into boxes, packed in sawdust, . . . trussed up in sacks, roped up like hams . . ." So similar in their treatment were the dead to ordinary items of commerce that every now and then boxes would be mixed up in transit. James Moores Ball, author of *The Sack-'Em-Up Men*, tells the tale of the flummoxed anatomist who opened a crate delivered to his lab expecting a cadaver but found instead "a very fine ham, a large cheese, a basket of eggs, and a huge ball of yarn." One can only imagine the surprise and very special disappointment of the party expecting very fine ham, cheese, eggs, or a huge ball of yarn, who found instead a well-packed but quite dead Englishman.

It wasn't so much the actual dissecting that smacked of disrespect. It was the whole street-theater-cum-abattoir air of the proceedings. Engravings by Thomas Rowlandson and William Hogarth of eighteenth- and early-nineteenth-century dissecting rooms show cadavers' intestines hanging like parade streamers off the sides of tables, skulls bobbing in boiling pots, organs strewn on the floor being eaten by dogs. In the background, crowds of men gawk and leer. While the artists were clearly editorializing upon the practice of dissection, written sources suggest the artworks were not far removed from the truth. Here is the composer Hector Berlioz, in an 1822 entry in his *Memoirs*, shedding considerable light on his decision to pursue music rather than medicine:

Robert . . . took me for the first time to the dissecting room. . . . At the sight of that terrible charnel-house—fragments of limbs, the grinning heads and gaping skulls, the bloody quagmire underfoot and the atrocious smell it gave off, the swarms of sparrows wrangling over scraps of lung,

the rats in their corner gnawing the bleeding vertebrae—
such a feeling of revulsion possessed me that I leapt through
the window of the dissecting room and fled for home as
though Death and all his hideous train were at my heels.

And I would wager a fine ham and a huge ball of yarn that
no anatomist of that era ever held a memorial service for the
leftover pieces. Cadaver remainders were buried not out of
respect but for lack of other options. The burials were hastily
done, always at night and usually out behind the building.

To avoid the problematic odors that tend to accompany a
shallow burial, anatomists came up with some creative solutions to the flesh disposal problem. A persistent rumor had
them in cahoots with the keepers of London's wild animal
menageries. Others were said to keep vultures on hand for the
task, though if Berlioz is to be believed, the sparrows of the
day were well up to the task. Richardson came across a reference to anatomists cooking down human bones and fat into "a
substance like Spermaceti," which they used to make candles
and soap. Whether these were used in the anatomists' homes
or given away as gifts was not noted, but between these and
the gastric-juice-etched nameplates, it's safe to say you really
didn't want your name on an anatomist's Christmas gift list.

And so it went. For nearly a century, the shortage of legally
dissectable bodies pitted the anatomist against the private citizen.
By and large, it was the poor who had most to lose. For over
time, entrepreneurs came up with an arsenal of antiresurrection-
ist products and services, affordable only by the upper class. Iron
cages called mortsafes could be set in concrete above the grave
or underground, around the coffin. Churches in Scotland built
graveyard "dead houses," locked buildings where a body could
be left to decompose until its structures and organs had disintegrated to the point where they were of no use to anatomists. You
could buy patented spring-closure coffins, coffins outfitted with

cast-iron corpse straps, double and even triple coffins. Appropriately, the anatomists were among the undertakers' best customers. Richardson relates that Sir Astley Cooper not only went for the triple coffin option but had the whole absurd Chinese-box affair housed in a hulking stone sarcophagus.

It was an Edinburgh anatomist named Robert Knox who instigated anatomy's fatal PR blunder: the implicit sanctioning of murder for medicine. In 1828, one of Knox's assistants answered the door to find a pair of strangers in the courtyard with a cadaver at their feet. This was business as usual for anatomists of the day, and so Knox invited the men in. Perhaps he made them a cup of tea, who knows. Knox was, like Astley, a man of high social bearing. Although the men, William Burke and William Hare, were strangers, he cheerfully bought the body and accepted their story that the cadaver's relatives had made the body available for sale—though this was, given the public's abhorrence of dissection, an unlikely scenario.

The body, it turns out, had been a lodger at a boardinghouse run by Hare and his wife, in an Edinburgh slum called Tanner's Close. The man died in one of Hare's beds, and, being dead, was unable to come up with the money he owed for the nights he'd stayed. Hare wasn't one to forgive a debt, so he came up with what he thought to be a fair solution: He and Burke would haul the body to one of those anatomists they'd heard about over at Surgeons' Square. There they would sell it, kindly giving the lodger the opportunity, in death, to pay off what he'd neglected to in life.

When Burke and Hare found out how much money could be made selling corpses, they set about creating some of their own. Several weeks later, a down-and-out alcoholic took ill with fever while staying at Hare's flophouse. Figuring the man to be well on his way to cadaverdom anyway, the men decided to speed things along. Hare pressed a pillow to the man's face while Burke laid his considerable body weight on top of him.

Knox asked no questions and encouraged the men to come back soon. And they did, some fifteen times. The pair were either too ignorant to realize that the same money could be made digging up graves of the already dead or too lazy to undertake it.

A series of modern-day Burke-and-Hare-type killings took place barely ten years ago, in Barranquilla, Colombia. The case centered on a garbage scavenger named Oscar Rafael Hernandez, who in March 1992 survived an attempt to murder him and sell his corpse to the local medical school as an anatomy lab specimen.* Like most of Colombia, Barranquilla lacked an organized recycling program, and hundreds of the city's destitute forge a living picking through garbage dumps for recyclables to sell. So scorned are these people that they— along with other social outcasts such as prostitutes and street urchins—are referred to as "disposables" and have often been murdered by right-wing "social cleansing" squads. As the story goes, guards from Universidad Libre had asked Hernandez if he wanted to come to the campus to collect some garbage, and then bludgeoned him over the head when he arrived. A *Los Angeles Times* account of the case has Hernandez awakening in a vat of formaldehyde alongside thirty corpses, a colorful if questionable detail omitted from other descriptions of the case. Either way, Hernandez came to and escaped to tell his tale.

Activist Juan Pablo Ordoñez investigated the case and claims that Hernandez was one of at least fourteen Barranquilla indigents murdered for medicine—even though an organized willed body program existed. According to Ordoñez's report, the national police had been unloading bodies gleaned from

---

* With the help of an interpreter, I got the number of an Oscar Rafael Hernandez living in Barranquilla. A woman answered the phone and said that Oscar was not in, whereupon my interpreter gamely asked her if Oscar was a garbage picker, and if he had been almost murdered by thugs who wanted to sell him to a medical school for dissection. A barrage of agitated Spanish ensued, which my interpreter summed up: "It's the wrong Oscar Rafael Hernandez."

their own, in-house "social cleansing" activities and collecting $150 per corpse from the university coffers. The school's security staff got wind of the setup and decided to get in on the action. At the time the investigation began, some fifty preserved bodies and body parts of questionable origin were found in the anatomy amphitheater. To date, no one from the university or the police has been arrested.

For his part, William Burke was eventually brought to justice. A crowd of more than 25,000 watched him hang. Hare was granted immunity, much to the disgust of the gallows crowd, who chanted "Burke Hare!"—meaning "Smother Hare," "burke" having made its way into the popular vernacular as a synonym for "smother." Hare probably did as much smothering as Burke, but "She's been hared!" lacks the pleasing Machiavellian fricatives of "She's been burked!" and the technicality is easily forgiven.

In a lovely sliver of poetic justice, Burke's corpse was, in keeping with the law of the day, dissected. As the lecture had been about the human brain, it seems unlikely that the body cavity would have been opened and notably rearranged, but perhaps this was thrown in after the fact, as a crowd pleaser. The following day the lab was opened to the public, and some thirty thousand vindicated gawkers filed past. The post-dissection cadaver was, by order of the judge, shipped to the Royal College of Surgeons of Edinburgh to have its bones made into a skeleton, which resides there to this day, along with one of several wallets made from Burke's skin.*

---

* Sheena Jones, the secretary at the college who told me about the wallet— which she called a "pocket book," nearly leading me to write that ladies' handbags had been made from Burke's hide—said it had been donated by one George Chiene, now deceased. Mrs. Jones did not know who had made or originally owned the wallet or whether Mr. Chiene had ever kept his money in it, but she observed that it looked like any other brown leather wallet and that "you would not know it is made from human skin."

Though Knox was never charged for his role in the murders, public sentiment held him accountable. The freshness of the bodies, the fact that one had its head and feet cut off and others had blood oozing from the nose or ears—all of this should have raised the bristly Knox eyebrows. The anatomist clearly didn't care. Knox further sullied his reputation by preserving one of Burke and Hare's more comely corpses, the prostitute Mary Paterson, in a clear glass vat of alcohol in his lab.

When an inquiry by a lay committee into Knox's role generated no formal action against the doctor, a mob gathered the following day with an effigy of Knox. (The thing must not have looked a great deal like the man, for they felt the need to label it: "Knox, the associate of the infamous Hare," explained a large sign on its back.) The stuffed Knox was paraded through the streets to the house of the real Knox, where it was hung by its neck from a tree and then cut down and—fittingly—torn to pieces.

It was around this time that Parliament conceded that the anatomy problem had gotten a tad out of hand and convened a committee to brainstorm solutions. While the debate mainly focused on alternate sources of bodies—most notably, unclaimed corpses from hospitals, prisons, and workhouses—some physicians raised an interesting item of debate: Is human dissection really necessary? Can't anatomy be learned from models, drawings, preserved prosections?

There have been times and places, in history, when the answer to the question "Is human dissection necessary?" was unequivocally yes. Here are some examples of what can happen when you try to figure out how a human body works without actually opening one up. In ancient China, Confucian doctrine considered dissection a defilement of the human body and forbade its practice. This posed a problem for the fathers of Chinese medicine, as is evident in this passage from the *Nei*

Ch'ing, or the *Canon of Medicine*, written around the tenth century. The fathers are rather clearly winging it:

> The heart is a king, who rules over all organs of the body; the lungs are his executive, who carry out his orders; the liver is his commandant, who keeps up the discipline; the gall bladder, his attorney general . . . and the spleen, his steward who supervises the five tastes. There are three burning spaces—the thorax, the abdomen and the pelvis—which are together responsible for the sewage system of the body.

Imperial Rome gives us another nice example of what happens to medicine when the government frowns on human dissection. Galen, one of history's most revered anatomists, whose texts went unchallenged for centuries, never once dissected a human cadaver. In his post as surgeon to the gladiators, he had a frequent, if piecemeal, window on the human interior in the form of gaping sword wounds and lion claw lacerations. He also dissected a good sum of animals, preferably apes, which he believed to be anatomically identical to humans, especially, he maintained, if the ape had a round face. The great Renaissance anatomist Vesalius later pointed out that there are two hundred anatomical differences between apes and humans in skeletal structure alone. (Whatever Galen's shortcomings as a comparative anatomist, the man is to be respected for his ingenuity, for procuring apes in ancient Rome can't have been easy.) He got a lot right, it's just that he also got a fair amount wrong. His drawings showed five-lobed livers and hearts with three ventricles.

The ancient Greeks were similarly adrift when it came to human anatomy. Like Galen, Hippocrates never dissected a human cadaver—he called dissection "unpleasant if not cruel." According to the book *Early History of Human Anatomy*, Hippocrates referred to tendons as "nerves" and believed the human brain to be a mucus-secreting gland. Though I found this information surprising, this being the Father of Medicine we are talking about, I did not question it. You do not question an author who appears on the title page as "T.V.N. Persaud, M.D., Ph.D., D.Sc., F.R.C.Path. (Lond.), F.F.Path. (R.C.P.I.), F.A.C.O.G." Who knows, perhaps history erred in bestowing upon Hippocrates the title Father of Medicine. Perhaps T.V.N. Persaud is the Father of Medicine.

It's no coincidence that the man who contributed the most to the study of human anatomy, the Belgian Andreas Vesalius, was an avid proponent of do-it-yourself, get-your-fussy-Renaissance-shirt-dirty anatomical dissection. Though human dissection was an accepted practice in the Renaissance-era anatomy class, most professors shied away from personally undertaking it, preferring to deliver their lectures while seated in raised chairs a safe and tidy remove from the corpse and pointing out structures with a wooden stick while a hired hand did the slicing. Vesalius disapproved of this practice, and wasn't shy about his feelings. In C. D. O'Malley's biography of the man, Vesalius likens the lecturers to "jackdaws aloft in their high chair, with egregious arrogance croaking things they have never investigated but merely committed to memory from the books of others. Thus everything is wrongly taught, . . . and days are wasted in ridiculous questions."

Vesalius was a dissector such as history had never seen. This was a man who encouraged his students to "observe the tendons while dining on any animal." While studying medicine in Belgium, he not only dissected the corpses of executed criminals but snatched them from the gibbet himself.

Vesalius produced a series of richly detailed anatomical plates and text called *De Humani Corporis Fabrica*, the most venerated anatomy book in history. The question then becomes, was it necessary, once the likes of Vesalius had pretty much figured out the basics, for every student of anatomy to get right in there and figure them out all over again? Why couldn't models and preserved prosections be used to teach anatomy? Do gross anatomy labs reinvent the wheel? The questions were especially relevant in Knox's day, given the way in which bodies were procured, but they are still relevant today.

I asked Hugh Patterson about this and learned that, in fact, whole-cadaver dissection is being phased out at some medical schools. Indeed, the gross anatomy course I visited at UCSF was the last one in which students will dissect entire cadavers. Beginning the following semester, they would be studying prosections—embalmed sections of the body cut and prepared so as to display key anatomical features and systems. Over at the University of Colorado, the Center for Human Simulation is leading the charge toward digital anatomy instruction. In 1993, they froze a cadaver and sanded off a millimeter cross section at a time, photographing each new view—1,871 in all—to create an on-screen, maneuverable 3-D rendition of the man and all his parts, a sort of flight simulator for students of anatomy and surgery.

The changes in the teaching of anatomy have nothing to do with cadaver shortages or public opinion about dissection; they have everything to do with time. Despite the immeasurable advances made in medicine over the past century, the material must be covered in the same number of years. Suffice it to say there's a lot less time for dissection than there was in Astley Cooper's day.

I asked the students in Patterson's gross anatomy lab how they'd feel if they hadn't had a chance to dissect a body. While some said they would feel cheated—that the gross anatomy cadaver experience was a physician's rite of passage—many expressed approval. "There were days," said one, "when it all clicked and I gained a sort of understanding I could never have gotten from a book. But there were other days, a lot of days, when coming up here and spending two hours felt like a huge waste of time."

But gross anatomy lab is not just about learning anatomy. It is about confronting death. Gross anatomy provides the medical student with what is very often his or her first exposure to a dead body; as such, it has long been considered a vital, necessary step in the doctor's education. But what was learned, up until quite recently, was not respect and sensitivity, but the opposite. The traditional gross anatomy lab represented a sort of sink-or-swim mentality about dealing with death. To cope with what was being asked of them, medical students had to find ways to desensitize themselves. They quickly learned to objectify cadavers, to think of the dead as structures and tissues, and not a former human being. Humor—at the cadaver's expense—was tolerated, condoned even. "There was a time not all that long ago," says Art Dalley, director of the Medical Anatomy Program at Vanderbilt University, "when students were taught to be insensitive, as a coping mechanism."

Modern educators feel there are better, more direct ways to address death than handing students a scalpel and assigning them a corpse. In Patterson's anatomy class at UCSF, as in many others, some of the time saved by eliminating full-body dissection will be devoted to a special unit on death and dying. If you're going to bring in an outsider to teach students about death, a hospice patient or grief counselor surely has as much to offer as a dead man does.

If the trend continues, medicine may find itself with something unimaginable two centuries ago: a surplus of cadavers. It is remarkable how deeply and how quickly public opinion regarding dissection and body donation has come around. I asked Art Dalley what accounted for the change. He cited a combination of factors. The 1960s saw the first heart transplant and the passing of the Uniform Anatomical Gift Act, both of which raised awareness of the need for organs for transplantation and of body donation as an option. Around the same time, Dalley says, there was a notable increase in the cost of funerals. This was followed by the publication of *The American Way of Death*—Jessica Mitford's biting exposé of the funeral industry—and a sudden upswing in the popularity of cremation. Willing one's body to science began to be seen as another acceptable—and, in this case, altruistic—alternative to burial.

To those factors I would add the popularization of science. The gains in the average person's understanding of biology have, I imagine, worked to dissolve the romance of death and burial—the lingering notion of the cadaver as some beatific being in an otherworldly realm of satin and chorale music, the well-groomed almost-human who simply likes to sleep a lot, underground, in his clothing. The people of the 1800s seemed to feel that burial culminated in a fate less ghastly than that of dissection. But that, as we'll see, is hardly the case.

3

LIFE AFTER DEATH

On human decay and
what can be done about it

Out behind the University of Tennessee Medical Center is a lovely, forested grove with squirrels leaping in the branches of hickory trees and birds calling and patches of green grass where people lie on their backs in the sun, or sometimes the shade, depending on where the researchers put them.

This pleasant Knoxville hillside is a field research facility, the only one in the world dedicated to the study of human decay. The people lying in the sun are dead. They are donated cadavers, helping, in their mute, fragrant way, to advance the science of criminal forensics. For the more you know about how dead bodies decay—the biological and chemical phases they go through, how long each phase lasts, how the environment affects these phases—the better equipped you are to figure out when any given body died: in other words, the day and even the approximate time of day it was murdered. The police are pretty good at pinpointing approximate time of death in recently dispatched bodies. The potassium level of the gel inside the eyes is helpful during the first twenty-four hours, as is algor mortis—the cooling of a dead body; barring temperature extremes, corpses lose about 1.5 degrees Fahrenheit per hour until they reach the temperature of the air around them. (Rigor mortis is more variable: It starts a few hours after death, usually in the head and neck, and continues, moving on down the body, finishing up and disappearing anywhere from ten to forty-eight hours after death.)

If a body has been dead longer than three days, investigators turn to entomological clues (e.g., how old are these fly larvae?) and stages of decay for their answers. And decay is highly dependent on environmental and situational factors. What's the weather been like? Was the body buried? In what? Seeking better understanding of the effects of these factors, the University of Tennessee (UT) Anthropological Research Facility, as it is blandly and vaguely called, has buried bodies in shallow graves, encased them in concrete, left them in car trunks and man-made ponds, and wrapped them in plastic bags. Pretty much anything a killer might do to dispose of a dead body the researchers at UT have done also.

To understand how these variables affect the time line of decomposition, you must be intimately acquainted with your control scenario: basic, unadulterated human decay. That's why I'm here. That's what I want to know: When you let nature take its course, just exactly what course does it take?

My guide to the world of human disassembly is a patient, amiable man named Arpad Vass. Vass has studied the science of human decomposition for more than a decade. He is an adjunct research professor of forensic anthropology at UT and a senior staff scientist at the nearby Oak Ridge National Laboratory. One of Arpad's projects at ORNL has been to develop a method of pinpointing time of death by analyzing tissue samples from the victim's organs and measuring the amounts of dozens of different time-dependent decay chemicals. This profile of decay chemicals is then matched against the typical profiles for that tissue for each passing postmortem hour. In test runs, Arpad's method has determined the time of death to within plus or minus twelve hours.

The samples he used to establish the various chemical

breakdown time lines came from bodies at the decay facility. Eighteen bodies, some seven hundred samples in all. It was an unspeakable task, particularly in the later stages of decomposition, and particularly for certain organs. "We'd have to roll the bodies over to get at the liver," recalls Arpad. The brain he got to using a probe through the eye orbit. Interestingly, neither of these activities was responsible for Arpad's closest brush with on-the-job regurgitation. "One day last summer," he says weakly, "I *inhaled* a fly. I could feel it buzzing down my throat."

I have asked Arpad what it's like to do this sort of work. "What do you mean?" he asked me back. "You want a vivid description of what's going through my brain as I'm cutting through a liver and all these larvae are spilling out all over me and juice pops out of the intestines?" I kind of did, but I kept quiet. He went on: "I don't really focus on that. I try to focus on the value of the work. It takes the edge off the grotesqueness." As for the humanness of his specimens, that no longer disturbs him. Though it once did. He used to lay the bodies on their stomachs so he didn't have to see their faces.

This morning, Arpad and I are riding in the back of a van being driven by the lovable and agreeable Ron Walli, one of ORNL's media relations guys. Ron pulls into a row of parking spaces at the far end of the UT Medical Center lot, labeled G section. On hot summer days, you can always find a parking space in G section, and not just because it's a longer walk to the hospital. G section is bordered by a tall wooden fence topped with concertina wire, and on the other side of the fence are the bodies. Arpad steps down from the van. "Smell's not that bad today," he says. His "not that bad" has that hollow, over-upbeat tone one hears when spouses back over flowerbeds or home hair coloring goes awry.

Ron, who began the trip in a chipper mood, happily pointing out landmarks and singing along with the radio, has the look of a condemned man. Arpad sticks his head in the window. "Are you coming in, Ron, or are you going to hide in the car again?" Ron steps out and glumly follows. Although this is his fourth time in, he says he'll never get used to it. It's not the fact that they're dead—Ron saw accident victims routinely in his former post as a newspaper reporter—it's the sights and smells of decay. "The smell just stays with you," he says. "Or that's what you imagine. I must have washed my hands and face twenty times after I got back from my first time out here."

Just inside the gate are two old-fashioned metal mailboxes on posts, as though some of the residents had managed to convince the postal service that death, like rain or sleet or hail, should not stay the regular delivery of the U.S. Mail. Arpad opens one and pulls turquoise rubber surgical gloves from a box, two for him and two for me. He knows not to offer them to Ron.

"Let's start over there." Arpad is pointing to a large male figure about twenty feet distant. From this distance, he could be napping, though there is something in the lay of the arms and the stillness of him that suggests something more permanent. We walk toward the man. Ron stays near the gate, feigning interest in the construction details of a toolshed.

Like many big-bellied people in Tennessee, the dead man is dressed for comfort. He wears gray sweatpants and a single-pocket white T-shirt. Arpad explains that one of the graduate students is studying the effects of clothing on the decay process. Normally, they are naked.

The cadaver in the sweatpants is the newest arrival. He will be our poster man for the first stage of human decay, the "fresh" stage. ("Fresh," as in fresh fish, not fresh air. As in recently dead but not necessarily something you want to put your nose right up to.) The hallmark of fresh-stage decay is a process called autolysis, or self-digestion. Human cells use enzymes to cleave molecules, breaking compounds down into things they can use. While a person is alive, the cells keep these enzymes in check, preventing them from breaking down the cells' own walls. After death, the enzymes operate unchecked and begin eating through the cell structure, allowing the liquid inside to leak out.

"See the skin on his fingertips there?" says Arpad. Two of the dead man's fingers are sheathed with what look like rubber fingertips of the sort worn by accountants and clerks. "The liquid from the cells gets between the layers of skin and loosens them. As that progresses, you see skin sloughage." Mortuary types have a different name for this. They call it "skin slip." Sometimes the skin of the entire hand will come off. Mortuary types don't have a name for this, but forensics types do. It's called "gloving."

"As the process progresses, you see giant sheets of skin peeling off the body," says Arpad. He pulls up the hem of the man's shirt to see if, indeed, giant sheets are peeling. They are not, and that's okay.

Something else is going on. Squirming grains of rice are crowded into the man's belly button. It's a rice grain mosh pit. But rice grains do not move. These cannot be grains of rice. They are not. They are young flies. Entomologists have a name for young flies, but it is an ugly name, an insult. Let's not use the word "maggot." Let's use a pretty word. Let's use "hacienda."

Arpad explains that the flies lay their eggs on the body's points of entry: the eyes, the mouth, open wounds, genitalia. Unlike older, larger haciendas, the little ones can't eat through skin. I make the mistake of asking Arpad what the little haciendas are after.

Arpad walks around to the corpse's left foot. It is bluish and the skin is transparent. "See the [haciendas] under the skin? They're eating the subcutaneous fat. They love fat." I see them.

They are spaced out, moving slowly. It's kind of beautiful, this man's skin with these tiny white slivers embedded just beneath its surface. It looks like expensive Japanese rice paper. You tell yourself these things.

Let us return to the decay scenario. The liquid that is leaking from the enzyme-ravaged cells is now making its way through the body. Soon enough it makes contact with the body's bacteria colonies: the ground troops of putrefaction. These bacteria were there in the living body as well, in the intestinal tract, in the lungs, on the skin—the places that came in contact with the outside world. Life is looking rosy for our one-celled friends. They've already been enjoying the benefits of a decommissioned human immune system, and now, suddenly, they're awash with this edible goo, issuing from the ruptured cells of the intestine lining. It's raining food. As will happen in times of plenty, the population swells. Some of the bacteria migrate to the far frontiers of the body, traveling by sea, afloat in the same liquid that keeps them nourished. Soon bacteria are everywhere. The scene is set for stage two: bloat.

The life of a bacterium is built around food. Bacteria don't have mouths or fingers or Wolf Ranges, but they eat. They digest. They excrete. Like us, they break their food down into its more elemental components. The enzymes in our stomachs break meat down into proteins. The bacteria in our gut break those proteins down into amino acids; they take up where we leave off. When we die, they stop feeding on what we've eaten and begin feeding on us. And, just as they do when we're alive, they produce gas in the process. Intestinal gas is a waste product of bacteria metabolism.

The difference is that when we're alive, we expel that gas. The dead, lacking workable stomach muscles and sphincters and bedmates to annoy, do not. Cannot. So the gas builds up and the belly bloats. I ask Arpad why the gas wouldn't just get forced out eventually. He explains that the small intestine has pretty much collapsed and sealed itself off. Or that there might be "something" blocking its egress. Though he allows, with some prodding, that a little bad air often does, in fact, slip out, and so, as a matter of record, it can be said that dead people fart. It needn't be, but it can.

Arpad motions me to follow him up the path. He knows where a good example of the bloat stage can be found.

Ron is still down by the shed, effecting some sort of gratuitous lawn mower maintenance, determined to avoid the sights and smells beyond the gate. I call for him to join me. I feel the need for company, someone else who doesn't see this sort of thing every day. Ron follows, looking at his sneakers. We pass a skeleton six feet seven inches tall and dressed in a red Harvard sweatshirt and sweatpants. Ron's eyes stay on his shoes. We pass a woman whose sizable breasts have decomposed, leaving only the skins, like flattened bota bags upon her chest. Ron's eyes stay on his shoes.

Bloat is most noticeable in the abdomen, Arpad is saying, where the largest numbers of bacteria are, but it happens in other bacterial hot spots, most notably the mouth and genitalia. "In the male, the penis and especially the testicles can become very large."

"Like how large?" (Forgive me.)

"I don't know. Large."

"Softball large? Watermelon large?"

"Okay, softball." Arpad Vass is a man with infinite reserves of patience, but we are scraping the bottom of the tank.

Arpad continues. Bacteria-generated gas bloats the lips and the tongue, the latter often to the point of making it protrude from the mouth: In real life as it is in cartoons. The eyes do not bloat because the liquid long ago leached out. They are gone. Xs. In real life as it is in cartoons.

Arpad stops and looks down. "That's bloat." Before us is a man with a torso greatly distended. It is of a circumference

I more readily associate with livestock. As for the groin, it is difficult to tell what's going on; insects cover the area, like something he is wearing. The face is similarly obscured. The larvae are two weeks older than their peers down the hill and much larger. Where before they had been grains of rice, here they are cooked rice. They live like rice, too, pressed together: a moist, solid entity. If you lower your head to within a foot or two of an infested corpse (and this I truly don't recommend), you can hear them feeding. Arpad pinpoints the sound: "Rice Krispies." Ron frowns. Ron used to like Rice Krispies.

Bloat continues until something gives way. Usually it is the intestines. Every now and then it is the torso itself. Arpad has never seen it, but he has heard it, twice. "A rending, ripping noise" is how he describes it. Bloat is typically short-lived, perhaps a week and it's over. The final stage, putrefaction and decay, lasts longest.

Putrefaction refers to the breaking down and gradual liquefaction of tissue by bacteria. It is going on during the bloat phase—for the gas that bloats a body is being created by the breakdown of tissue—but its effects are not yet obvious.

Arpad continues up the wooded slope. "This woman over here is farther along," he says. That's a nice way to say it. Dead people, unembalmed ones anyway, basically dissolve; they collapse and sink in upon themselves and eventually seep out onto the ground. Do you recall the Margaret Hamilton death scene in *The Wizard of Oz*? ("I'm melting!") Putrefaction is more or less a slowed-down version of this. The woman lies in a mud of her own making. Her torso appears sunken, its organs gone—leached out onto the ground around her.

The digestive organs and the lungs disintegrate first, for they are home to the greatest numbers of bacteria; the larger your work crew, the faster the building comes down. The brain is another early-departure organ. "Because all the bacteria in the mouth chew through the palate," explains Arpad. And because brains are soft and easy to eat. "The brain liquefies very quickly. It just pours out the ears and bubbles out the mouth."

Up until about three weeks, Arpad says, remnants of organs can still be identified. "After that, it becomes like a soup in there." Because he knew I was going to ask, Arpad adds, "Chicken soup. It's yellow."

Ron turns on his heels. "Great." We ruined Rice Krispies for Ron, and now we have ruined chicken soup.

Muscles are eaten not only by bacteria, but by carnivorous beetles. I wasn't aware that meat-eating beetles existed, but there you go. Sometimes the skin gets eaten, sometimes not. Sometimes, depending on the weather, it dries out and mummifies, whereupon it is too tough for just about anyone's taste. On our way out, Arpad shows us a skeleton with mummified skin, lying facedown. The skin has remained on the legs as far as the tops of the ankles. The torso, likewise, is covered, about up to the shoulder blades. The edge of the skin is curved, giving the appearance of a scooped neckline, as on a dancer's leotard. Though naked, he seems dressed. The outfit is not as colorful or, perhaps, warm as a Harvard sweatsuit, but more fitting for the venue.

We stand for a minute, looking at the man.

There is a passage in the Buddhist Sutra on Mindfulness called the Nine Cemetery Contemplations. Apprentice monks are instructed to meditate on a series of decomposing bodies in the charnel ground, starting with a body "swollen and blue and festering," progressing to one "being eaten by . . . different kinds of worms," and moving on to a skeleton, "without flesh and blood, held together by the tendons." The monks were told to keep meditating until they were calm and a smile appeared on their faces. I describe this to Arpad and Ron, explaining that the idea is to come to peace with the transient nature of our bodily existence, to overcome the revulsion and fear. Or something.

We all stare at the man. Arpad swats at flies.

"So," says Ron. "Lunch?"

Outside the gate, we spend a long time scraping the bottoms of our boots on a curb. You don't have to step on a body to carry the smells of death with you on your shoes. For reasons we have just seen, the soil around a corpse is sodden with the liquids of human decay. By analyzing the chemicals in this soil, people like Arpad can tell if a body has been moved from where it decayed. If the unique volatile fatty acids and compounds of human decay aren't there, the body didn't decompose there.

One of Arpad's graduate students, Jennifer Love, has been working on an aroma scan technology for estimating time of death. Based on a technology used in the food and wine industries, the device, now being funded by the FBI, would be a sort of hand-held electronic nose that could be waved over a body and used to identify the unique odor signature that a corpse puts off at different stages of decay.

I tell them that the Ford Motor Company developed an electronic nose programmed to identify acceptable "new car smell." Car buyers expect their purchases to smell a certain way: leathery and new, but with no vinyl off-gassy smells. The nose makes sure the cars comply. Arpad observes that the new-car-smell electronic nose probably uses a technology similar to what the electronic nose for cadavers would use.

"Just don't get 'em confused," deadpans Ron. He is imagining a young couple, back from a test drive, the woman turning to her husband and saying: "You know, that car smelled like a dead person."

It is difficult to put words to the smell of decomposing human. It is dense and cloying, sweet but not flower-sweet. Halfway between rotting fruit and rotting meat. On my walk home each afternoon, I pass a fetid little produce store that gets the mix almost right, so much so that I find myself peering behind the papaya bins for an arm or a glimpse of naked feet. Barring a visit to my neighborhood, I would direct the curious to a chemical supply company, from which one can order synthetic versions of many of these volatiles. Arpad's lab has rows of labeled glass vials: Skatole, Indole, Putrescine, Cadaverine. The moment wherein I uncorked the putrescine in his office may well be the moment he began looking forward to my departure. Even if you've never been around a decaying body, you've smelled putrescine. Decaying fish throws off putrescine, a fact I learned from a gripping *Journal of Food Science* article entitled "Post-Mortem Changes in Black Skipjack Muscle During Storage in Ice." This fits in with something Arpad told me. He said he knew a company that manufactured a putrescine detector, which doctors could use in place of swabs and cultures to diagnose vaginitis or, I suppose, a job at the skipjack cannery.

The market for synthetic putrescine and cadaverine is small, but devoted. The handlers of "human remains dogs" use these compounds for training.* Human remains dogs are distinct from the dogs that search for escaped felons and the dogs that search for whole cadavers. They are trained to alert their owners when they detect the specific scents of decomposed human tissue. They can pinpoint the location of a corpse at the bottom of a lake by sniffing the water's surface for the gases and fats

---

* Purists among them insist on the real deal. I spent an afternoon in an abandoned dormitory at Moffett Federal Airfield, watching one such woman, Shirley Hammond, put her canine noses through their paces. Hammond is a fixture on the base, regularly seen walking to and from her car with a pink gym bag and a plastic cooler. If you were to ask her what she's got in there, and she chose to answer you honestly, the answer would go more or less like this: a bloody shirt, dirt from beneath a decomposed corpse, human tissue buried in a chunk of cement, a piece of cloth rubbed on cadavers, a human molar. No synthetics for Shirley's dogs.

that float up from the rotting remains. They can detect the lingering scent molecules of a decomposing body up to fourteen months after the killer lugged it away.

I had trouble believing this when I heard it. I no longer have trouble. The soles of my boots, despite washing and soaking in Clorox, would smell of corpse for months after my visit.

Ron drives us and our little cloud of stink to a riverside restaurant for lunch. The hostess is young and pink and clean-looking. Her plump forearms and tight-fitting skin are miracles. I imagine her smelling of talcum powder and shampoo, the light, happy smells of the living. We stand apart from the hostess and the other customers, as though we were traveling with an ill-tempered, unpredictable dog. Arpad signals to the hostess that we are three. Four, if you count The Smell.

"Would you like to sit indoors . . . ?"

Arpad cuts her off. "Outdoors. And away from people."

That is the story of human decay. I would wager that if the good people of the eighteenth and nineteenth centuries had known what happens to dead bodies in the sort of detail that you and I now know, dissection might not have seemed so uniquely horrific. Once you've seen bodies dissected, and once you've seen them decomposing, the former doesn't seem so dreadful. Yes, the people of the eighteenth and nineteenth centuries were buried, but that only served to draw out the process. Even in a coffin six feet deep, the body eventually decomposes. Not all the bacteria living in a human body require oxygen; there are plenty of anaerobic bacteria up to the task.

Nowadays, of course, we have embalming. Does this mean we are spared the unsavory fate of gradual liquefaction? Has modern mortuary science created an eternity free from unpleasant mess and stains? Can the dead be aesthetically pleasing? Let's go see!

An eye cap is a simple ten-cent piece of plastic. It is slightly larger than a contact lens, less flexible, and considerably less comfortable. The plastic is repeatedly lanced through, so that small, sharp spurs stick up from its surface. The spurs work on the same principle as those steel spikes that threaten Severe Tire Damage on behalf of rental car companies: The eyelid will come down over an eye cap, but, once closed, will not easily open back up. Eye caps were invented by a mortician to help dead people keep their eyes shut.

There have been times this morning when I wished that someone had outfitted me with a pair of eye caps. I've been standing around, eyelids up, in the basement embalming room of the San Francisco College of Mortuary Science.

Upstairs is a working mortuary, and above it are the classrooms and offices of the college, one of the nation's oldest and best-respected.* In exchange for a price break in the cost of embalming and other mortuary services, customers agree to let students practice on their loved ones. Like getting a $5 haircut at the Vidal Sassoon Academy, sort of, sort of not.

I had called the college to get answers about embalming: How long does it preserve corpses, and in what form? Is it possible to never decompose? How does it work? They agreed to answer my questions, and then they asked me one. Did I want to come down and see how it's done? I did, sort of, sort of not.

Presiding at the embalming table today are final-semester students Theo Martinez and Nicole D'Ambrogio. Theo, a

---

* And, alas, most expensive and least well attended. In May 2002, a year after I visited, it closed its doors.

dark-haired man of thirty-nine with a long, distinguished face and narrow build, turned to mortuary science after a string of jobs in credit unions and travel agencies. He says he liked the fact that mortuary jobs often include housing. (Before cell phones and pagers, most funeral homes were built with apartments, so that someone was always there should a call come in at night.) For the beautiful and glossy-haired Nicole, episodes of *Quincy* sparked an interest in the career, which is a little puzzling, because Quincy, if I recall, was a pathologist. (No matter what they say, the answer never quite satisfies.) The pair are garbed in plastic and latex, as am I and anyone else who plans to enter the "splash area." They are working with blood; the garments are a precaution against it and all it may bring on: HIV, hepatitis, stains on your shirt.

The object of their attentions at the moment is a seventy-five-year-old man, or a three-week-old cadaver, however you prefer to think of it. The man had donated his body to science, but, owing to its having been autopsied, science politely declined. An anatomy lab is as choosy as a pedigreed woman seeking love: You can't be too fat or too tall or have any communicable diseases. Following a three-week sojourn in a university refrigerator, the cadaver wound up here. I have agreed to disguise any identifying features, though I suspect that the dehydrating air of refrigeration has gotten a jump on the task. He looks gaunt and desiccated. Something of the old parsnip about him.

Before the embalming begins, the exterior of the corpse is cleaned and groomed, as it would be were this man to be displayed in an open casket or presented to the family for a private viewing. (In reality, when the students are through, no one but the cremation furnace attendant will see him.) Nicole swabs the mouth and eyes with disinfectant, then rinses both with a jet of water. Though I know the man to be dead, I expect to see him flinch when the cotton swab hits his eye, to cough and sputter when the water hits the back of his throat. His stillness, his deadness, is surreal.

The students move purposefully. Nicole is looking in the man's mouth. Her hand rests sweetly on his chest. Concerned, she calls Theo over to look. They talk quietly and then he turns to me. "There's material sitting in the mouth," he says.

I nod, picturing corduroy, swatches of gingham. "Material?"

"Purge," offers Nicole. It's not helping.

Hugh "Mack" McMonigle, an instructor at the college, who is supervising this morning's session, steps up beside me. "What happened is that whatever was in the stomach found its way into the mouth." Gases created by bacterial decay build up and put pressure on the stomach, squeezing its contents back up the esophagus and into the mouth. The situation appears not to bother Theo and Nicole, though purge is a relatively infrequent visitor to the embalming room.

Theo explains that he is going to use an aspirator. As if to distract me from what I am seeing, he keeps up a friendly patter. "The Spanish for 'vacuum' is *aspiradora*."

Before switching on the aspirator, Theo takes a cloth to the man's chin and wipes away a substance that looks but surely doesn't taste like chocolate syrup. I ask him how he copes with the unpleasantnesses of dealing with dead strangers' bodies and secretions. Like Arpad Vass, he says that he tries to focus on the positives. "If there are parasites or the person has dirty teeth or they didn't wipe their nose before they died, you're improving the situation, making them more presentable."

Theo is single. I ask him whether studying to be a mortician has been having a deleterious effect on his love life. He straightens up and looks at me. "I'm short, I'm thin, I'm not rich. I would say my career choice is in fourth place in limiting my effectiveness as a single adult." (It's possible that it helped. Within a year, he would be married.)

Next Theo coats the face with what I assume to be some

sort of disinfecting lotion, which looks a lot like shaving cream. The reason that it looks a lot like shaving cream, it turns out, is that it is. Theo slides a new blade into a razor. "When you shave a decedent, it's really different."

"I bet."

"The skin isn't able to heal, so you have to be really careful about nicks. One shave per razor, and then you throw it away." I wonder whether the man, in his dying days, ever stood before a mirror, razor in hand, wondering if it might be his last shave, unaware of the actual last shave that fate had arranged for him.

"Now we're going to set the features," says Theo. He lifts one of the man's eyelids and packs tufts of cotton underneath to fill out the lid the way the man's eyeballs once did. Oddly, the culture I associate most closely with cotton, the Egyptians, did not use their famous Egyptian cotton for plumping out withered eyes. The ancient Egyptians put pearl onions in there. *Onions.* Speaking for myself, if I had to have a small round martini garnish inserted under my eyelids, I would go with olives.

On top of the cotton go a pair of eye caps. "People would find it disturbing to find the eyes open," explains Theo, and then he slides down the lids. In the corner of my viewing screen, my brain displays a special pull-out graphic, an animated close-up of the little spurs in action. *Madre de dio! Aspiradora!* Come the day, you won't be seeing me in an open casket.

As a feature of the common man's funeral, the open casket is a relatively recent development: around 150 years. According to Mack, it serves several purposes, aside from providing what undertakers call "the memory picture." It reassures the family that, one, their loved one is unequivocally dead and not about to be buried alive, and, two, that the body in the casket is indeed their loved one, and not the stiff from the container beside his. I read in *The Principles and Practice of Embalming* that it came into vogue as a way for embalmers to show off their skills. Mack disagrees, noting that long before embalming became commonplace, corpses on ice inside their caskets were displayed at funerals. (I am inclined to believe Mack, this being a book that includes the passage "Many of the body tissues also possess some measure of immortality if they can be kept under proper conditions. . . . Theoretically, it is possible in this way to grow a chicken heart to the size of the world.")

"Did you already go in the nose?" Nicole is holding aloft tiny chrome scissors. Theo says no. She goes in, first to trim the hair, then with the disinfectant. "It gives the decedent some dignity," she says, plunging wadded cotton into and out of his left nostril.

I like the term "decedent." It's as though the man weren't dead, but merely involved in some sort of protracted legal dispute. For evident reasons, mortuary science is awash with euphemisms. "Don't say stiff, corpse, cadaver," scolds *The Principles and Practice of Embalming*. "Say decedent, remains or Mr. Blank. Don't say 'keep.' Say 'maintain preservation.' . . ." Wrinkles are "acquired facial markings." Decomposed brain that filters down through a damaged skull and bubbles out the nose is "frothy purge."

The last feature to be posed is the mouth, which will hang open if not held shut. Theo is narrating for Nicole, who is using a curved needle and heavy-duty string to suture the jaws together. "The goal is to reenter through the same hole and come in behind the teeth," says Theo. "Now she's coming out one of the nostrils, across the septum, and then she's going to reenter the mouth. There are a variety of ways of closing the mouth," he adds, and then he begins talking about something called a needle injector. I pose my own mouth to resemble the mouth of someone who is quietly horrified, and this works quite well to close Theo's mouth. The suturing proceeds in silence.

Theo and Nicole step back and regard their work. Mack nods. Mr. Blank is ready for embalming.

Modern embalming makes use of the circulatory system to deliver a liquid preservative to the body's cells to halt autolysis and put decay on hold. Just as blood in the vessels and capillaries once delivered oxygen and nutrients to the cells, now those same vessels, emptied of blood, are delivering embalming fluid. The first people known to attempt arterial embalming* were a trio of Dutch biologists and anatomists named Swammerdam, Ruysch, and Blanchard, who lived in the late 1600s. The early anatomists were dealing with a chronic shortage of bodies for dissection, and consequently were motivated to come up with ways to preserve the ones they managed to obtain. Blanchard's textbook was the first to cover arterial embalming. He describes opening up an artery, flushing the blood out with water, and pumping in alcohol. I've been to frat parties like that.

Arterial embalming didn't begin to catch on in earnest until the American Civil War. Up until this point, dead U.S. soldiers were buried more or less where they fell. Their families had to send a written request for disinterment and ship a coffin capable of being hermetically sealed to the nearest quartermaster office, whereupon the quartermaster officer would assign a team of men to dig up the remains and deliver them to the family. Often the coffins that the families sent were not

---

* But by no means the first to attempt to keep bodies from rotting. Outtakes of the early days of corporeal preservation included a seventeenth-century Italian physician named Girolamo Segato, who devised a way of turning bodies into stone, and a London M.D. named Thomas Marshall, who, in 1839, published a paper describing an embalming technique that entailed generously puncturing the surface of the body with scissors and then brushing the body with vinegar, much the way the Adolph's company would have housewives prick steaks to get the meat tenderizer way down in.

hermetically sealed—who knew what "hermetically" meant? Who knows now?—and they soon began to stink and leak. At the urgent pleadings of the beleaguered delivery brigades, the army set about embalming its dead, some 35,000 in all.

One fine day in 1861, a twenty-four-year-old colonel named Elmer Ellsworth was shot and killed as he seized a Confederate flag from atop a hotel, his rank and courage a testimony to the motivating powers of a humiliating first name. The colonel was given a hero's send-off and a first-class embalming at the hands of one Thomas Holmes, the Father of Embalming.* The public filed past Elmer in his casket, looking every bit the soldier and nothing at all the decomposing body. Embalming received another boost four years later, when Abe Lincoln's embalmed body traveled from Washington to his hometown in Illinois. The train ride amounted to a promotional tour for funerary embalming, for wherever the train stopped, people came to view him, and more than a few must have noted that he looked a whole lot better in his casket than Grandmama had looked in hers. Word spread and the practice grew, like a chicken heart, and soon the whole nation was sending their decedents in to be posed and preserved.

After the war, Holmes set up a business selling his patented embalming fluid, Innominata, to embalmers, but otherwise began to distance himself from the mortuary trade. He opened a drugstore, manufactured root beer, and invested in a health spa, and between the three of them managed to squander his considerable savings. He never married and fathered no children (other than Embalming), but it wouldn't be accurate to say he lived alone. According to Christine Quigley, author of

---

* Does everything have a father? Apparently so. A web search on "the father of" turned up fathers for vasectomy reversal, hillbilly jazz, lichenology, snowmobiling, modern librarianship, Japanese whiskey, hypnosis, Pakistan, natural hair care products, the lobotomy, women's boxing, Modern Option Pricing Theory, the swamp buggy, Pennsylvania ornithology, Wisconsin bluegrass, tornado research, Fen-Phen, modern dairying, Canada's permissive society, black power, and the yellow schoolbus.

The Corpse: A History, he shared his Brooklyn house with samples of his war-era handiwork: Embalmed bodies were stored in the closets, and heads sat on tables in the living room. Not all that surprisingly, Holmes began to go insane, spending his final years in and out of institutions. At seventy, he was placing ads in mortuary trade journals for a rubber-coated canvas body removal bag that could, he suggested, *double as a sleeping bag*. Shortly before he died, Holmes is said to have requested that he not be embalmed, though whether this was a function of sanity or insanity was never made clear.

Theo is feeling around on Mr. Blank's neck. "We're in search of the carotid artery," he announces. He cuts a short lengthwise slit in the man's neck. Because no blood flows, it is easy to watch, easy to think of the action as simply something a man does on his job, like cutting roofing material or slicing foam core, rather than what it would more normally be: murder. Now the neck has a secret pocket, and Theo slips his finger into it. After some probing, he finds and raises the artery, which is then severed with a blade. The loose end is pink and rubbery and looks very much like what you blow into to inflate a whoopee cushion.

A cannula is inserted into the artery and connected by a length of tubing to the canister of embalming fluid. Mack starts the pump.

Here is where it all begins to make sense. Within minutes, the man's face looks rejuvenated. The embalming fluid has rehydrated his tissues, filling out his sunken cheeks, his lined skin. His skin is pink now (the embalming fluid contains red coloring), no longer slack and papery. He looks healthy and surprisingly alive. This is why you don't just stick bodies in the refrigerator before an open-casket funeral.

Mack is telling me about a ninety-seven-year-old woman

who looked sixty after her embalming. "We had to paint in wrinkles, or the family wouldn't recognize her."

As hale and youthful as our Mr. Blank looks this morning, he will still eventually decompose. Mortuary embalming is designed to keep a cadaver looking fresh and uncadaverous for the funeral service, but not much longer. (Anatomy departments amp up the process by using greater amounts and higher concentrations of formalin; these corpses may remain intact for years, though they take on a kind of pickled horror-movie appearance.) "As soon as the water table comes up, and the coffin gets wet," Mack allows, "you're going to have the same kind of decomposition you would have had if you hadn't done embalming." Water reverses the chemical reactions of embalming, he says.

Funeral homes sell sealed vaults designed to keep air and water out, but even then, the corpse's prospects for eternal comeliness are iffy. The body may contain bacteria spores, hardy suspended-animation DNA pods, able to withstand extremes of temperature, dryness, and chemical abuse, including that of embalming. Eventually the formaldehyde breaks down, and the coast is clear for the spores to bring forth bacteria.

"Undertakers used to claim embalming was permanent," says Mack. "If it meant making the sale on that family, believe me, that embalmer was going to say anything," agrees Thomas Chambers, of the W. W. Chambers chain of funeral homes, whose grandfather walked the boundaries of taste when he distributed promotional calendars featuring a nude silhouette of a shapely woman above the mortuary's slogan, "Beautiful Bodies by Chambers." (The woman was not, as Jessica Mitford seemed to hint in *The American Way of Death,* a cadaver that the mortuary had embalmed; that would have been going too far, even for Grandpa Chambers.)

Embalming fluid companies used to encourage experimentation by sponsoring best-preserved-body contests. The hope

was that some undertaker, by craft or serendipity, would figure
out the perfect balance of preservatives and hydrators, enabling
his trade to preserve a body for years without mummifying it.
Contestants were invited to submit photographs of decedents
who had held up particularly well, along with a write-up of
their formulas and methods. The winning entries and photos
would be published in mortuary trade journals, on the pre–
Jessica Mitford assumption that no one outside the business
ever cracked an issue of *Casket and Sunnyside*.

I asked Mack what made the undertakers back off from their
claims of eternal preservation. It was, as it so often is, a lawsuit.
"One man took them up on it. He bought a space in a mauso-
leum and every six months he'd go in with his lunch and open
up his mother's casket and visit with her on his lunch hour.
One especially wet spring, some moisture got in, and come to
find, Mom had grown a beard. She was covered with mold.
He sued, and collected twenty-five thousand dollars from
the mortuary. So they've stopped making that statement."
Further discouragement has come from the Federal Trade
Commission, whose 1982 Funeral Rule prohibited mortuary
professionals from claiming that the coffins they sold provided
eternal protection against decay.

And that is embalming. It will make a good-looking corpse
of you for your funeral, but it will not keep you from one day
dissolving and reeking, from becoming a Halloween ghoul. It
is a temporary preservative, like the nitrites in your sausages.
Eventually any meat, regardless of what you do to it, will
wither and go off.

The point is that no matter what you choose to do with your
body when you die, it won't, ultimately, be very appealing. If
you are inclined to donate yourself to science, you should not
let images of dissection or dismemberment put you off. They
are no more or less gruesome, in my opinion, than ordinary
decay or the sewing shut of your jaws via your nostrils for a

funeral viewing. Even cremation, when you get right down to it—as W.E.D. Evans, former Senior Lecturer in Morbid Anatomy at the University of London, did in his 1963 book *The Chemistry of Death*—isn't a pretty event:

The skin and hair at once scorch, char and burn. Heat coagulation of muscle protein may become evident at this stage, causing the muscles slowly to contract, and there may be a steady divarication of the thighs with gradually developing flexion of the limbs. There is a popular idea that early in the cremation process the heat causes the trunk to flex forwards violently so that the body suddenly "sits up," bursting open the lid of the coffin, but this has not been observed personally. . . .

Occasionally there is swelling of the abdomen before the skin and abdominal muscles char and split; the swelling is due to formation of steam and the expansion of gases in the abdominal contents.

Destruction of the soft tissues gradually exposes parts of the skeleton. The skull is soon devoid of covering, then the bones of the limbs appear. . . . The abdominal contents burn fairly slowly, and the lungs more slowly still. It has been observed that the brain is specially resistant to complete combustion during cremation of the body. Even when the vault of the skull has broken and fallen away, the brain has been seen as a dark, fused mass with a rather sticky consistency. . . . Eventually the spine becomes visible as the viscera disappear, the bones glow whitely in the flames and the skeleton falls apart.

Drops of sweat bead the inside surface of Nicole's splash shield. We've been here more than an hour. It's almost over. Theo looks at Mack. "Will we be suturing the anus?" He turns to me. "Otherwise leakage can wick into the funeral clothing and it's an awful mess."

I don't mind Theo's matter-of-factness. Life contains these things: leakage and wickage and discharge, pus and snot and slime and gleet. We are biology. We are reminded of this at the beginning and the end, at birth and at death. In between we do what we can to forget.

Since our decedent will not be having a funeral service, it is up to Mack whether the students must take the final step. He decides to let it go. Unless the visitor wishes to see it. They look at me.

"No thank you." Enough biology for today.

4

DEAD MAN DRIVING

———

Human crash test dummies and the ghastly,
necessary science of impact tolerance

By and large, the dead aren't very talented. They can't play water polo, or lace up their boots, or maximize market share. They can't tell a joke, and they can't dance for beans. There is one thing dead people excel at. They're very good at handling pain.

For instance, UM 006. UM 006 is a cadaver who recently journeyed across Detroit from the University of Michigan to the bioengineering building at Wayne State University. His job, which he will undertake at approximately 7 P.M. tonight, is to be hit in the shoulder with a linear impactor. His collarbone and scapula may break, but he will not feel a thing, nor will the injuries interfere with his day-to-day activities. By agreeing to be walloped in the shoulder, cadaver UM 006 is helping researchers figure out how much force a human shoulder in a side-impact car crash can withstand before it registers a serious injury.

Over the past sixty years, the dead have helped the living work out human tolerance limits for skull slammings and chest skewerings, knee crammings and gut mashings: all the ugly, violent things that happen to a human being in a car crash. Once automobile manufacturers know how much force a skull or spine or shoulder can withstand, they can design cars that, they hope, will not exceed that force in a crash.

You are perhaps wondering, as I did, why they don't use crash test dummies. This is the other side of the equation. A dummy can tell you how much force a crash is unleashing on various dummy body parts, but without knowing how much of a blow a real body part can take, the information is useless. You first need to know, for instance, that the maximum amount a rib cage can compress without damaging the soft, wet things inside it is 2¾ inches. Then, should a dummy slam into a steering wheel of a newly designed car and register a chest deflection of four inches, you know the National Highway Traffic Safety Administration (NHTSA) isn't going to be very happy with that car.

The dead's first contribution to safe driving was the non-face-gashing windshield. The first Fords came without windshields, which is why you see pictures of early motorists wearing goggles. They weren't trying to affect a dashing World War I flying-ace mien; they were keeping wind and bugs out of their eyes. The first windscreens were made of ordinary window glass, which served to cut the wind and, unfortunately, the driver's face in the event of a crash. Even with the early laminated-glass windshields, which were in use from the 1930s to the mid-1960s, front-seat passengers were walking away from accidents with gruesome, gaping scalp-to-chin lacerations. Heads would hit the windshield, knock out a head-shaped hole in the glass, and, on their violent, bouncing return back through that hole, get sliced open on the jagged edges.

Tempered glass, the follow-up innovation, was strong enough to keep heads from smashing through, but the concern then became that striking the stiffer glass would cause brain damage. (The less a material gives, the more damaging the forces of the impact: Think ice rink versus lawn.) Neurologists knew that a concussion from a forehead impact was accompanied by some degree of skull fracture. You can't give a dead man a concussion, but you can check his skull for hairline cracks, and this is what researchers did. At Wayne State, cadavers were leaned forward over a simulated car window and dropped from varying heights (simulating varying speeds) so that their foreheads hit the glass. (Contrary to popular impression, impact test cadavers were not typically ushered into the front seats of actual running automobiles, driving being one of the other things cadavers don't do well. More often than not, the cadaver was either dropped or it remained still while some sort of controllable impacting device was directed at it.) The study showed that tempered glass, provided it wasn't too thick, was unlikely to create forces strong enough to cause concussion. Windshields today have even more give, enabling the modern-day head to undergo a 30-mph unbelted car crash straight into a wall and come away with little to complain about save a welt and an owner whose driving skills are up there with the average cadaver's.

Despite forgiving windshields and knobless, padded dashboards, brain damage is still the major culprit in car crash fatalities. Very often, the bang to the head isn't all that severe. It's the combination of banging it into something and whipping it in one direction and then rapidly back at high speeds (rotation, this is called) that tends to cause serious brain damage. "If you hit the head without any rotation, it takes a huge amount of force to knock you out," says Wayne State Bioengineering Center director Albert King. "Similarly, if you rotate the head without hitting anything, it's hard to cause severe damage." (High-speed rear-enders sometimes do this; the brain is whipped back and forth so fast that shear forces tear open the veins on its surface.) "In the run-of-the-mill crash, there's some of each, neither of which is very high, but you can get a severe head injury." The sideways jarring of a side-impact crash is especially notorious for putting passengers in comas.

King and some of his colleagues are trying to get a handle on what, exactly, is happening to the brain in these banging/

whipping-around scenarios. Across town at Henry Ford Hospital, the team has been filming cadavers' heads with a high-speed X-ray video camera* during simulated crashes, to find out what's going on inside the skull. So far they're finding a lot more "sloshing of the brain," as King put it, with more rotation than was previously thought to occur. "The brain traces out a kind of figure eight," says King. It is something best left to skaters: When brains do this they get what's called diffuse axonal injury—potentially fatal tears and leaks in the microtubules of the brain's axons.

Chest injuries are the other generous contributor to crash fatalities. (This was true even before the dawn of the automobile; the great anatomist Vesalius, in 1557, described the burst aorta of a man thrown from his horse.) In the days before seat belts, the steering wheel was the most lethal item in a car's interior. In a head-on collision, the body would slide forward and the chest would slam into the steering wheel, often with enough force to fold the rim of the wheel around the column, in the manner of a closing umbrella. "We had a guy take a tree head-on and there was the N from the steering wheel—the car was a Nash—imprinted in the center of his chest," recalls Don Huelke, a safety researcher who spent the years from 1961 through 1970 visiting the scene of every car accident fatality in the county surrounding the University of Michigan and recording what happened and how.

Steering wheel columns up through the sixties were narrow,

---

* Other lively things to do with X-ray video cameras: At Cornell University, bio-mechanics researcher Diane Kelley has filmed lab rats mating in X-ray, in order to shed light on the possible role of the penis bone. Humans do not have penis bones, nor have they, to the author's knowledge, been captured having sex on X-ray videotape. They have, however, been filmed having sex inside an MRI tube, by fun-loving physiologists at the University Hospital in Groningen, Netherlands. The researchers concluded that during intercourse in the missionary position, the penis "has the shape of a boomerang."

sometimes only six or seven inches in diameter. Just as a ski pole will sink into the snow without its circular basket, a steering column with its rim flattened back will sink into a body. In an unfortunate design decision, the steering wheel shaft of the average automobile was angled and positioned to point straight at the driver's heart.* In a head-on, you'd be impaled in pretty much the last place you'd want to be impaled. Even when the metal didn't penetrate the chest, the impact alone was often fatal. Despite its thickness, an aorta ruptures relatively easily. This is because every other second, it has a one-pound weight suspended from it: the human heart, filled with blood. Get the weight moving with enough force, as happened in blunt impacts from steering wheels, and even the body's largest blood vessel can't take the strain. If you insist on driving around in vintage cars with no seat belt on, try to time your crashes for the systole—blood-squeezed-out—portion of your heartbeat.

With all this in mind, bioengineers and automobile manufacturers (GM, notably) began ushering cadavers into the driver's seats of crash simulators, front halves of cars on machine-accelerated sleds that are stopped abruptly to mimic the forces of a head-on collision. The goal, one of them anyway, was to design a steering column that would collapse on impact, absorbing enough of the shock to prevent serious injury to the heart and its supporting vessels. (Hoods are now designed to do this too, so that even cars in relatively minor

---

* From a safety standpoint, it would have been better to skip steering wheels entirely and install a pair of rudderlike handles on either side of the driver's seat, as was done in the "Survival Car," a traveling demo car built by the Liberty Mutual Insurance Company in the early 1960s to show the world how to build cars that save lives (and reduce insurance company payouts). Other visionary design elements included a rear-facing front passenger seat, a feature about as likely to sell cars as, well, steering rudders. Safety did not sell automobiles in the sixties, style did, and the Survival Car failed to change the world.

accidents have completely jackknifed hoods, the idea being
that the more the car crumples, the less you do.) GM's first
collapsible steering wheel shaft, introduced in the early 1960s,
cut the risk of death in a head-on collision by half.

And so it went. The collective cadaver résumé boasts contributions to government legislation for lap-shoulder belts,
air bags, dashboard padding, and recessed dashboard knobs
(autopsy files from the 1950s and 1960s contain more than a
few X-ray images of human heads with radio knobs embedded in them). It was not pretty work. In countless seat-belt
studies—car manufacturers, seeking to save money, spent
years trying to prove that seat belts caused more injuries than
they prevented and thus shouldn't be required—bodies were
strapped in and crashed, and their innards were then probed
for ruptures and manglings. To establish the tolerance limits of
the human face, cadavers have been seated with their cheekbones in the firing lines of "rotary strikers." They've had their
lower legs broken by simulated bumpers and their upper legs
shattered by smashed-in dashboards.

It is not pretty, but it is most certainly justifiable. Because of
changes that have come about as a result of cadaver studies, it's
now possible to survive a head-on crash into a wall at 60 mph.
In a 1995 *Journal of Trauma* article entitled "Humanitarian Benefits of Cadaver Research on Injury Prevention," Albert King
calculated that vehicle safety improvements that have come
about as a result of cadaver research have saved an estimated
8,500 lives each year since 1987. For every cadaver that rode
the crash sleds to test three-point seat belts, 61 lives per year
have been saved. For every cadaver that took an air bag in the
face, 147 people per year survive otherwise fatal head-ons. For
every corpse whose head has hammered a windshield, 68 lives
per year are saved.

Unfortunately, King did not have these figures handy in
1978, when chairman John Moss of the House Subcommittee

on Oversight and Investigations called a hearing to investigate the use of human cadavers in car crash testing. Representative Moss said he felt a "personal repugnance about this practice." He said that there had developed within NHTSA "a sort of cult that finds that this is a necessary tool." He believed that there had to be another way to go about it. He wanted proof that dead people in crashing cars behave exactly like living ones—proof that, as exasperated researchers pointed out, could never be obtained because it would mean subjecting a series of live humans to exactly the same high-force impacts as a series of dead humans.

Oddly, Representative Moss was not a squeamish man when it came to dead bodies; he had worked briefly in a funeral parlor before he entered politics. Nor was he an especially conservative man. He was a Democrat, a pro-safety reformer. What had got him agitated, said King (who testified at the hearing), was this: He had been working to pass legislation to make air bags mandatory and was infuriated by a cadaver test that showed an air bag causing more injury than a seat belt. (Air bags sometimes do injure, even kill, particularly if the passenger is leaning forward or otherwise OOP—"out of position"— but in this case, to be fair to Moss, the air bag body was older and probably frailer.) Moss was an oddity: an automotive safety lobbier taking a stand against cadaver research.

In the end, with the support of the National Academy of Sciences, the Georgetown Center for Bioethics, the National Catholic Conference, a chairman of a noted medical school's anatomy department who stated that "such experiments are probably as highly respectful [as medical school anatomy dissections] and less destructive to the human body," and representatives of the Quaker, Hindu, and Reform Judaism religions, the committee concluded that Moss himself was a tad "out of position." There is no better stand-in for a live human in a car crash than a dead one.

Lord knows, the alternatives have been tried. In the dawn of impact science, researchers would experiment on themselves. Albert King's predecessor at the Bioengineering Center, Lawrence Patrick, volunteered himself as a human crash test dummy for years. He has ridden the crash sled some four hundred times, and been slammed in the chest by a twenty-two-pound metal pendulum. He has hurled one knee repeatedly against a metal bar outfitted with a load cell. Some of Patrick's students were equally courageous, if courageous is the word. A 1965 Patrick paper on knee impacts reports that student volunteers seated in crash sleds endured knee impacts equivalent to a force of one thousand pounds. The injury threshold was estimated at fourteen hundred pounds. His 1963 study "Facial Injuries—Cause and Prevention" includes a photograph of a young man who appears to be resting peacefully with his eyes shut. Closer inspection hints that, in fact, something not at all peaceful is about to unfold. For starters, the man is using a book entitled *Head Injuries* as a headrest (uncomfortable, but probably pleasanter than reading it). Hovering just above the man's cheek is a forbidding metal rod identified in the caption as a "gravity impactor." The text informs us that "the volunteer waited several days for the swelling to subside and then the test was continued up to the energy limit which he could endure." Here was the problem. Impact data that doesn't exceed the injury threshold is of minimal use. You need those folks who don't feel pain. You need cadavers.

Moss wanted to know why animals couldn't be used in automotive impact testing, and indeed they have been. A description of the Eighth Stapp Car Crash and Field Demonstration Conference, which appears in the introduction to its proceedings, begins like a child's recollections of a trip to the circus: "We saw chimpanzees riding rocket sleds, a bear on an impact swing. . . . We observed a pig, anaesthetized and placed in a sitting position on the swing in the harness, crashed into a deep-dished steering wheel. . . ."

Pigs were popular subjects because of their similarities to humans "in terms of their organ setup," as one industry insider put it, and because they can be coaxed into a useful approximation of a human sitting in a car. As far as I can tell, they are also similar to a human sitting in a car in terms of their intelligence setup, their manners setup, and pretty much everything else, excluding possibly their use of cupholders and ability to work the radio buttons, but that is neither here nor there. In more recent years, animals have typically been used only when functioning organs are needed, and cadavers cannot oblige. Baboons, for example, have been subjected to violent sideways head rotations in order to study why side-impact crashes so often send passengers into comas. (Researchers, in turn, were subject to violent animal rights protests.) Live dogs were recruited to study aortic rupture; for unknown reasons, it has proved difficult to experimentally rupture a cadaver aorta.

There is one type of automotive impact study in which animals are still used even though cadavers would be vastly more accurate, and that is the pediatric impact study. No child donates his remains to science, and no researcher wants to bring up body donation with grieving parents, even though the need for data on children and air-bag injuries has been obvious and dire. "It's a real problem," Albert King told me. "We try to scale it from baboons, but the strength is all different. And a kid's skull is not completely formed; it changes as it grows." In 1993, a research team at the Heidelberg University School of Medicine had the courage to attempt a series of impact studies on children—and the audacity to do it without consent. The press got hold of it, the clergy got involved, and the facility was shut down.

Child data aside, the blunt impact tolerance limits of the

human body's vital pieces have long ago been worked out, and today's dead are being recruited mainly for impact studies of the body's outlying regions: ankles, knees, feet, shoulders. "In the old days," King told me, "people involved in severe crashes ended up in the morgue." No one cares about a dead man's shattered ankle. "Now these guys are surviving because of the air bag, and we have to worry about these things. You have people with both ankles and knees damaged and they will never walk right again. It's a major disability now."

Tonight at Wayne State's impact lab, a cadaver shoulder impact is taking place, and King has been gracious enough to invite me to watch. Actually, he didn't invite me. I asked if I could watch, and he agreed to it. Still, considering what I'll be seeing and how sensitive the public is to these things and further considering that Albert King has read my writing and knows it doesn't exactly read like *The International Journal of Crashworthiness,* he was pretty darn gracious.

Wayne State has been involved in impact research since 1939, longer than any other university. On the wall above the landing of the front stairs of the Bioengineering Center a banner proclaims: "Celebrating 50 Years of Moving Forward with Impact." It is 2001, which suggests that for twelve years now, no one has thought to take down the banner, which you kind of expect from engineers.

King is on his way to the airport, so he leaves me with fellow bioengineering professor John Cavanaugh, who will be overseeing tonight's impact. Cavanaugh looks at once like an engineer and a young Jon Voight, if that's possible. He has a laboratory complexion, pale and unlined, and regular-looking brown hair. When he talks or shifts his glance, his eyebrows rise and his forehead draws together, giving him a more or less permanent look of mild worry. Cavanaugh brings me downstairs to the impact lab. It is a typical university lab, with ancient, jerry-rigged equipment and decor that runs to block-lettered safety reminders. Cavanaugh introduces me to Matt Mason, tonight's research assistant, and Deb Marth, for whose Ph.D. dissertation the impact is being done, and then he disappears upstairs.

I glance around the room for UM 006, the way, as a child, I used to scan the basement for the thing that reaches through the banisters to grab your legs. He isn't here yet. A crash test dummy sits on a sled railing. Its upper body rests on its thighs, head on knees, as though collapsed in despair. It has no arms, perhaps the source of the despair.

Matt is linking up high-speed videocameras to a pair of computers and to the linear impactor. The impactor is a formidably sized piston fired by compressed air and mounted on a steel base the size of a fairground pony. From the hallway, a sound of clattering wheels. "Here he comes," says Deb. UM 006 lies on a gurney being wheeled by a muscular man with gray hair and rambunctious eyebrows, dressed, like Marth, in surgical scrubs.

"I am Ruhan," says the man beneath the eyebrows. "I am the cadaver man." He holds out a gloved hand. I wave, to show him that I'm not wearing gloves. Ruhan comes from Turkey, where he was a doctor. For a former doctor whose job now entails diapering and dressing cadavers, he has an admirably upbeat disposition. I ask him if it's difficult to dress a dead man, and how he does it. Ruhan describes the process, then stops. "Have you ever been to a nursing home? It's like that."

UM 006 is dressed this evening in a Smurf-blue leotard and matching tights. Beneath the tights he wears a diaper, for leakage. The neckline of his leotard is wide and scooped, like a dancer's. Ruhan confirms that the cadaver leotards are purchased from a dancers' supply house. "They would be disgusted if they knew!" To ensure anonymity, the dead man's face is masked by a snug-fitting white cotton hood. He looks like someone about to rob a bank, someone who meant to pull pantyhose over his head but got it wrong and used an athletic sock.

Matt sets down his laptop and helps Ruhan lift the cadaver into the car seat, which sits on a table beside the impactor. Ruhan is right. It's nursing-home work: dressing, lifting, arranging. The distance between the very old, sick, frail person and the dead one is short, with a poorly marked border. The more time you spend with the invalid elderly (I have seen both my parents in this state), the more you come to see extreme old age as a gradual easing into death. The old and the dying sleep more and more, until one day they "sleep" all the time. They often become more and more immobile until one day they can do no more than lie or sit however the last person positioned them. They have as much in common with UM 006 as they do with you and me.

I find the dead easier to be around than the dying. They are not in pain, not afraid of death. There are no awkward silences and conversations that dance around the obvious. They aren't scary. The half hour I spent with my mother as a dead person was easier by far than the many hours I spent with her as a live person dying and in pain. Not that I wished her dead. I'm just saying it's easier. Cadavers, once you get used to them—and you do that quite fast—are surprisingly easy to be around.

Which is good, because at the moment, it's just he and I. Matt is in the next room, Deb has gone to look for something. UM 006 was a big, meaty man, still is. His tights are lightly stained. His leotard shows up his lumpy, fallen midsection. The aging superhero who can't be bothered to wash his costume. His hands are mittened with the same cotton as his head. It was probably done to depersonalize him, as is done with the hands of anatomy lab cadavers, but for me it has the opposite effect. It makes him seem vulnerable and toddlerlike.

Ten minutes pass. Sharing a room with a cadaver is only mildly different from being in a room alone. They are the same sort of company as people across from you on subways or in airport lounges, there but not there. Your eyes keep going back to them, for lack of anything more interesting to look at, and then you feel bad for staring.

Deb is back. She is checking accelerometers that she has painstakingly mounted to exposed areas of the cadaver's bones: on the scapula, clavicle, vertebrae, sternum, and head. By measuring how fast the body accelerates on impact, the devices essentially give you the force of the hit, as measured in g's (gravities). After the test, Deb will autopsy the shoulder area and catalog the damage at this particular speed. What she is after is the injury threshold and the forces needed to generate it; the information will be used to develop shoulder instrumentation for the SID, the side-impact dummy.

A side-impact accident is one in which the cars collide at ninety degrees, bumper to door, the kind that often take place at four-way intersections when one party hasn't bothered to stop at the light or heed the stop sign. Lap-shoulder belts and dashboard air bags are engineered to protect against the forward-heaving forces of a head-on crash; they do little for a person in a side-impact crash. The other thing working against you in this type of crash is the immediacy of the other car; there is no engine or trunk and rear seat to absorb the blow.* There are a couple inches of metal door. This is also the reason it took so long for side air bags to begin appearing in cars. With no hood to collapse, the sensors have to sense the impact immediately, and the old ones weren't up to the task.

---

* This is why you shouldn't worry all that much about sitting in the middle seat, without a shoulder belt. If the car gets hit from the side, you're better off being farther from the doors. The kindly people beside you, the ones with the shoulder belts, will absorb the impact for you.

Deb knows all about this because she works as a design engineer at Ford and was the person who implemented the side air
bags in the 1998 Town Car. She doesn't look like an engineer.
She has magazine-model skin and a wide, white, radiant smile
and thick, shiny brown hair pulled back in a loose ponytail.
If Julia Roberts and Sandra Bullock had a child together, it
would look like Deb Marth.

The cadaver before UM 006 was hit at a faster speed: 15
mph (which, were this a real side-impact accident with a passenger door to absorb some of the energy of the impact, would
translate to being hit by a car going perhaps 25 or 30 mph).
The impact broke his collarbone and scapula and fractured
five ribs. Ribs are more important than you think. When you
breathe, you not only need to move your diaphragm to pull
air into your lungs, you need the muscles attached to your ribs
and the ribs themselves. If all your ribs break, your rib cage
can't help inflate your lungs the way it's supposed to, and you
will find it very hard to breathe. It is a condition called "flail
chest," and people die from it.

Flail chest is one of the other things that make side impacts
especially dangerous. Ribs are easier to break from the side.
The rib cage is built to be compressed from the front, sternum
to spine—that's how it moves when you breathe. (Up to a
point, that is. Compress it too far and you can, in the words of
Don Huelke, "split the heart completely in half as you would
a pear.") A rib cage is not built for the sideways press. Slam it
violently from the side, and its tines tend to snap.

Matt is still working on the setup. Deb is intent on her accelerometers. Normally, accelerometers are screwed into place,
but if she were to screw them into the bone, the bone would
be weakened and would break more easily in the impact.
Instead she secures them to the bone with wire ties and then
wedges wood shims underneath to tighten the fit. As she
works, she slips the wire cutters into and out of the cadaver's

mittened hand, as though he were a surgical nurse. Another way for him to help.

With the radio playing and the three of us talking, the room has a feeling of late-night congeniality. I find myself thinking that it's nice for UM 006 to have company. There can be no lonelier state of being than that of being a corpse. Here, in the lab, he's part of something, part of the group, the center of everyone's attention. Of course, these are stupid thoughts, for UM 006 is a mass of tissue and bone who can no more feel loneliness than he can feel Marth's fingers probing the flesh around his collarbone. But that's how I feel about it at the moment.

It is past nine now. UM 006 has begun to put out a subtle gamy smell, the mild but unmistakable fetor of a butcher shop on a hot afternoon. "How long," I ask, "can he stay out at room temperature before he starts to . . ." Marth waits for me to finish my sentence. " . . . change?" She says maybe half a day. She is looking put-upon. The ties aren't tight enough and the Krazy Glue's not crazy anymore. It's going to be a long night.

John Cavanaugh calls down that there's pizza upstairs, and the three of us, Deb, Matt Mason, and I, leave the dead man by himself. It feels a little rude.

On the way upstairs, I ask Deb how she wound up working with dead bodies for a living. "Oh, I always wanted to do cadaver research," she says, with exactly the same enthusiasm and sincerity with which a more usual individual would say "I always wanted to be an archaeologist" or "I always wanted to live by the sea."

"John was so psyched. Nobody wants to do cadaver research." In her office, she takes a bottle of a perfume called Happy from a desk drawer. "So I smell something else," she explains. She has promised to give me some papers, and while she searches for them I look at a pile of snapshots on her desk. And then, very quickly, I don't. The photographs are close-ups from a previous cadaver's shoulder autopsy: meaty red and parted skin. Matt looks down at the pile. "These aren't your vacation shots, are they, Deb?"

By half past eleven, all that remains is to get UM 006 into driving posture. He is slumped and leaning to one side. He is the guy next to you on the plane, asleep and inching closer to your shoulder.

John Cavanaugh takes the cadaver by the ankles and pushes back on him, to try to get him to sit up in the seat. He steps back. The cadaver slides back toward him. He pushes him again. This time he holds him while Matt encircles UM 006's knees and the entire circumference of the car seat with duct tape. "This probably won't make it into the '101 Uses' list," observes Matt.

"His head's wrong," says John. "It needs to be straight ahead." More duct tape. The radio is playing the Romantics, "That's What I Like About You."

"He's slumping again."

"Try the winch?" Deb loops a canvas strap under his arms and presses a button that raises a ceiling-mounted motor winch. The cadaver shrugs, slowly, and holds it, like a Borscht Belt comedian. He lifts slightly from his seat, and is lowered back down, sitting straighter now. "Good, perfect," says John.

Everyone steps back. UM 006 has a comic's timing. He waits a beat, two beats, then slips forward again. You have to laugh. The absurdity of the scene and the punch-drunk hour are making it hard not to. Deb gets some pieces of foam to prop up his back, which seems to do the trick.

Matt runs a final check of the connections. The radio—I'm not making this up—is playing "Hit Me with Your Best Shot." Five more minutes pass. Matt fires the piston. It sounds a loud bang as it shoots out, though the impact itself is silent. UM 006 falls over, not like a villain shot in a Hollywood movie, but slowly, like an off-balance laundry sack. He falls over onto a foam pad that has been set out for this purpose, and John and Deb step forward to steady him. And that's that. Without the screech of skidding tires and the crunch and fold of metal, an impact is neither violent nor disturbing. Distilled to its essence, controlled and planned, it is now simply science, no longer tragedy.

The family of UM 006 does not know what happened to him this evening. They know only that he donated his remains for use in medical education or research. There are many reasons for this. At the time a person or his family decides to donate his remains, no one knows what those remains will be used for, or even at which university. The body goes to a morgue facility at the university to which it was donated, but may be shipped, as was UM 006, from that school to another.

For a family to be fully informed of what is happening to their loved one, the information would have to come from the researchers themselves, after they've taken receipt of the body (or body part) but before they run their test. As a result of the subcommittee hearings, that was sometimes done. Automotive impact researchers who received federal NHTSA funding and who had not made it clear in their willed body consent forms that the remains might be used for research were required to contact families prior to the experiment. According to Rolf Eppinger, chief of the NHTSA Biomechanics Research Center, it was rare for the family to renege on the deceased's consent.

I spoke with Mike Walsh, who worked for one of NHTSA's main contractors, Calspan. It was Walsh who, once the body arrived, called the family to set up a meeting—preferably, owing to the highly perishable state of unembalmed remains, within a day or two after the death. You would think, as principal investigator on these studies, that Walsh would have delegated the enormously uncomfortable task to someone else. But Walsh preferred to do it himself. He told the families precisely how their loved one would be used and why. "The entire program was explained to them. Some studies were sled impact studies, some were pedestrian impact studies,* some were in full-scale crash vehicles." Clearly Walsh has a gift. Out of forty-two families contacted, only two revoked consent—not because of the nature or specifics of the study, but because they had thought the body was going to be used for organ donation.

I asked Walsh whether any family members had asked to see a copy of the study when it was published. No one had. "We got the impression, quite frankly, that we were giving people more information than they wanted to hear."

In England and other Commonwealth countries, researchers and anatomy instructors sidestep the possibility of family or public disapproval by using body parts and prosections—the name given to embalmed cadaver segments used in anatomy labs—rather than whole cadavers. England's antivivisectionists, as animal rights activists are called there, are as outspoken as America's, and the things that outrage them are more encompassing, and, dare I say it, nonsensical. To give you a taste: In 1916, a group of animal rights activists successfully petitioned the British Undertakers Association on behalf of

---

* To quote a Stapp Car Crash Conference study on the topic, "Pedestrians are not 'run over' by cars. They are 'run under.'" It typically goes like this: Bumper hits calf and front of hood hits hip, knocking the legs out from under and flipping them up over the head. The cartwheeling pedestrian then lands on his head or chest on the hood or windshield. Depending on the speed of the impact, he may continue cartwheeling, legs over head again, and land flat on the roof, and from there slide off onto the pavement. Or he may remain on the hood, with his head smashed through the windshield. Whereupon the driver calls an ambulance, unless the driver is someone like Fort Worth nurse's aide Chante Mallard, in which case she keeps on driving, returns to her house, and allegedly leaves the car in the garage with the victim sticking out of her windshield until he bleeds to death. This event took place in October 2001. Mallard was arrested and charged with murder.

the horses that pulled their hearses, urging members to stop making the horses wear plumes on their heads.

The British investigators know what butchers have long known: If you want people to feel comfortable about dead bodies, cut them into pieces. A cow carcass is upsetting; a brisket is dinner. A human leg has no face, no eyes, no hands that once held babies or stroked a lover's cheek. It's difficult to associate it with the living person from which it came. The anonymity of body parts facilitates the necessary dissociations of cadaveric research: This is not a person. This is just tissue. It has no feelings, and no one has feelings for it. It's okay to do things to it which, were it a sentient being, would constitute torture.

But let's be rational. Why is it okay for someone to guide a table saw through Granddad's thigh and then pack up the leg for shipment to a lab, where it will be suspended from a hook and impacted with a simulated car bumper, yet not okay to ship him and use him whole? What makes cutting his leg off first any less distasteful or disrespectful? In 1901, the French surgeon René Le Fort devoted a great deal of his time to studying the effects of blunt impact on the bones of the face. Sometimes he severed the heads: "After decapitation, the head was violently thrown against the rounded border of a marble table . . . ," reads an experiment description from *The Maxillo-Facial Works of René Le Fort.* Other times he left the heads on: "The entire cadaver was in a dorsal . . . position with the head hanging back over the table. A violent blow was given with a wooden club on the right upper jaw. . . ." What person who takes offense at the latter could reasonably be comfortable with the former? What, ethically or aesthetically, is the difference?

Furthermore, it's often desirable, from the standpoint of biomechanical fidelity, to use the entire enchilada. A shoulder mounted on a stand and hit with an impactor doesn't behave in the same manner, or incur the same injuries, as a shoulder mounted on a torso. When shoulders on stands start getting driver's licenses, then it will make sense to study them. Even scientific inquiries as seemingly straightforward as *How much will a human stomach hold before it bursts?* have gone the extra mile. In 1891, an inquiring German doctor surnamed Key-Aberg undertook a replication of a French study done six years earlier, in which isolated human stomachs were filled to the point of rupture. Key-Aberg's experiment differed from that of his French predecessor in that he left the stomachs inside their owners. He presumably felt that this better approximated the realities of a hearty meal, for rare indeed is the dinner party attended by freestanding stomachs. To that end, he is said to have made a point of composing his corpses in the sitting position. In this case, our man's attention to biomechanical correctness proved not to matter. In both cases, according to a 1979 article in *The American Journal of Surgery,* the stomachs gave out at 4,000 cc's, or about four quarts.*

---

* As fans of the eating sections of old Guinness books of world records will surmise, this figure has been surpassed on numerous occasions. Some stomachs, by way of heredity or prolonged daily gourmandism, are roomier than average. Orson Welles's was one such stomach. According to the owners of Pink's hot dog stand in L.A., the voluminous director once sat down and finished off eighteen franks.

The all-time record holder would appear to be a twenty-three-year-old London fashion model whose case was described in the April 1985 *Lancet.* At what turned out to be her last meal, the young woman managed to put away nineteen pounds of food: one pound of liver, two pounds of kidney, a half pound of steak, one pound of cheese, two eggs, two thick slices of bread, one cauliflower, ten peaches, four pears, two apples, four bananas, two pounds each of plums, carrots, and grapes, and two glasses of milk. Whereupon her stomach blew and she died. (The human gastrointestinal tract is home to trillions of bacteria, which, should they escape the confines of their stinky, labyrinthine home, create a massive and often fatal systemic infection.)

Runner-up goes to a thirty-one-year-old Florida psychologist who was found collapsed in her kitchen. The Dade County medical examiner's report itemized the fatal last meal: "8700 cc of poorly masticated, undigested hot dogs, broccoli and cereal suspended in a green liquid that contained numerous small bubbles." The green liquid remains a mystery, as does the apparent widespread appeal of hot dogs among modern-day gorgers (from Salon.com).

Many times, of course, a researcher doesn't need a whole body, just a piece of it. Orthopedic surgeons developing new techniques or new replacement joints use limbs instead of whole cadavers. Ditto product safety researchers. You do not need an entire dead body to find out, say, what happens to a finger when you close a particular brand of power window on it. You need some fingers. You do not need an entire body to see whether softer baseballs cause less damage to Little Leaguers' eyes. You need some eyes, mounted in clear plastic simulated eye sockets so that high-speed video cameras can document exactly what is happening when the baseballs hit them.*

Here's the thing: No one really *wants* to work with whole cadavers. Unless researchers really need to, they won't. Rather than use whole bodies to simulate swimmers in a test of a safety cage for outboard motor propellers, Tyler Kress, who runs the Sports Biomechanics Lab at the University of Tennessee's Engineering Institute for Trauma and Injury Prevention, went to the trouble of tracking down artificial ball-and-socket hip joints and gluing them to cadaver legs with surgical cement and then gluing the resulting cadaver-leg-and-hip-joint hybrid to a crash test dummy torso.

Kress says it wasn't fear of public reprisal that led him to do this, but practicality. "A leg," he told me, "is so much easier

---

* This was a subject of heated debate in ophthalmology corners. Some felt that if you made baseballs softer, they would deform on impact and penetrate more deeply into the socket, causing more damage, not less. A study done by researchers at the Vision Performance and Safety Service at Tufts University School of Medicine showed that softer balls did indeed penetrate more deeply, but they didn't cause more damage. That would have been tough to do, for the harder balls ruptured the eye "from the limbus to the optic nerve with almost total extrusion of the intraocular contents." Let us hope that the manufacturers of amateur sports equipment have read the March 1999 *Archives of Ophthalmology* and adjusted the hardness of their baseballs accordingly. Either way, eye protection for Little Leaguers is a swell idea.

to work with and handle." Parts are easier to lift and maneuver. They take up less space in the freezer. Kress has worked with just about all of them: heads, spines, shins, hands, fingers. "Legs, mostly," he says. He spent last summer looking at the biomechanics of twisted and broken ankles. This summer he and his colleagues are running instrumented leg-drop tests to look at the sorts of injuries that accompany vertical drops, such as befall mountain bikers and snowboarders. "I would challenge you to find anybody that's broken more legs than we have."

I asked Kress, in an e-mail exchange, whether he has had occasion to wrangle a cadaveric crotch into an athletic cup and take aim at it with baseballs, hockey pucks, what-have-you. He has not, nor is he aware of any sports injury researcher who has. "You would think that . . . 'racking'—i.e., scrotal impacts—would be a high research priority," he wrote. "I'm thinking no one wants to go there in the lab."

Which is not to say that science does not, occasionally, go there. At the local medical school library, I ran a Pub Med search for journal articles featuring the words "cadaveric" and "penis." With the monitor shoved back as far as possible into the cubicle, lest the people on either side of me see the screen and alert the librarian, I browsed twenty-five entries, most of them anatomical investigations. There were the urologists from Seattle investigating the distribution pattern of dorsal nerves along the penile shaft (twenty-eight cadaver penises).[*] There were the French anatomists injecting red liquid latex into penile arteries to study vascular flow (twenty cadaver penises). There were the Belgians calculating interference of

---

[*] This was a joint effort involving the living and the dead, with the dead getting the shorter end of the stick: Following dissections of the dead penises, "10 healthy males" agreed to help confirm the findings by undergoing electrical stimulation of the dorsal nerve, as healthy males are wont to agree to.

the ischiocavernosus muscles in rigidity during penile erection (thirty cadaver penises). For the past twenty years, all the world over, people in white coats and squeaking shoes have been calmly, methodically making the cut that dare not speak its name. It makes Tyler Kress seem like a cream puff.

On the other side of the gender gap, a Pub Med search on "clitoris" and "cadaver" turned up but a single entry. Australian urologist Helen O'Connell, author of "Anatomical Relationship Between Urethra and Clitoris" (ten cadaver perinea), bristles at the disparity: "Modern anatomy texts," she writes, "have reduced descriptions of female perineal anatomy to a brief adjunct after a complete description of the male anatomy." I picture O'Connell as a sort of Gloria Steinem of the research set, the fast-moving, can-do feminist in a lab coat. She is also the first researcher I've come across in my haphazard wanderings to have worked with infant cadavers. (She did this because the comparable male erectile tissue research had, for reasons not explained, been done on infants.) Her paper states that she obtained ethical approval from the Victorian Institute of Forensic Pathology and the Board of Medical Research of the Royal Melbourne Hospital, which clearly don't go about their business with the grim specter of media evisceration foremost in their minds.

5

BEYOND THE BLACK BOX

When the bodies of the passengers
must tell the story of a crash

Dennis Shanahan works in a roomy suite on the second floor
of the house he shares with his wife, Maureen, in a subdivision
ten minutes east of downtown Carlsbad, California. The office
is quiet and sunny and offers no hint of the grisly nature of
the work done within. Shanahan is an injury analyst. Much of
the time, he analyzes the wounds and breakages of the living.
He consults for car companies being sued by people making
dubious claims ("the seat belt broke," "I wasn't driving," and so
on) that are easily debunked by looking at their injuries. Every
now and then the bodies he studies are dead ones. Such was
the case with TWA Flight 800.

Bound for Paris from JFK International Airport on July 17,
1996, Flight 800 blew apart in the air over the Atlantic off
East Moriches, New York. Witness reports were contradictory.
Some claimed to have seen a missile strike the aircraft. Traces
of explosives had turned up in the recovered wreckage, but no
trace of bomb hardware had been found. (Later it would come
out that the explosive materials had been planted in the plane
long before the crash, as part of a sniffer-dog training exercise.) Conspiracy theories sprouted and spread. The investigation dragged on without a definitive answer to the question

on everyone's mind: What—or who—had brought Flight 800 down from the sky?

Within days of the crash, Shanahan flew to New York to visit the bodies of the dead and see what they had to say. Last spring, I flew to Carlsbad, California, to visit Shanahan. I wanted to know how—scientifically and emotionally—a person does this job.

I had other questions for him too. Shanahan is a man who knows the reality behind the nightmare. He knows, in grim medical detail, exactly what happens to people in different types of crashes. He knows how they typically die, whether they're likely to have been cognizant of what was happening, and how—in a low-altitude crash, anyway—they might have increased their chances of survival. I told him I would only take up an hour of his time, but stayed for five.

A crashed plane will usually tell its own story. Sometimes literally, in the voices on the cockpit flight recorder; sometimes by implication, in the rendings and charrings of the fallen craft. But when a plane goes down over the ocean, its story may be patchy and incoherent. If the water is especially deep or the currents swift and chaotic, the black box may not be recovered, nor may enough of the sunken wreckage be recovered to determine for sure what occurred in the plane's last minutes. When this happens, investigators turn to what is known in aviation pathology textbooks as "the human wreckage": the bodies of passengers. For unlike a wing or a piece of fuselage, a corpse will float to the water's surface. By studying victims' wounds—the type, the severity, which side of the body they're on—an injury analyst can begin to piece together the horrible unfolding of events.

Shanahan is waiting for me when I arrive at the airport. He is wearing Dockers, a short-sleeved shirt, and aviator-frame glasses. His hair lies neatly on either side of a perfectly straight part. It could almost be a toupee, but isn't. He is polite, composed, and immediately likable. He reminds me of my pharmacist Mike.

He isn't at all what I'd had in mind. I had imagined someone gruff, morgue-hardened, prone to expletives. I had planned to do my interview in the field, in the aftermath of a crash. I pictured the two of us in a makeshift morgue in some small-town dance hall or high-school gym, he in his stained lab coat, me with my notepad. This was before I realized that Shanahan himself does not do the autopsies for the crashes he investigates. These are done by teams of medical examiners from nearby county morgues. Though he goes to the site and will often examine bodies for one reason or another, Shanahan works mostly with the autopsy reports, correlating these with the flight's seating chart to identify clusters of telltale injuries. He explained that visiting him at work on a crash site might have required a wait of several years, for the cause of most crashes isn't a mystery, and thus input from the cadavers isn't often called for.

When I tell him I was disappointed over not being able to report from the scene of a crash, Shanahan hands me a book called *Aerospace Pathology,* which, he assures me, contains photographs of the sorts of things I might have seen. I open the volume to a section on "body plotting." Among line sketches of downed plane pieces, small black dots are scattered. Leader lines spoke away from the dots to their labels: "brown leather shoes," "copilot," "piece of spine," "stewardess." By the time I get to the chapter that describes Shanahan's work—"Patterns of Injury in Fatal Aircraft Accidents," wherein photo captions remind investigators to keep in mind things like "intense heat may produce intracranial steam resulting in blowout of the cranial vault, simulating injuries from impact"—it has become clear to me that labeled black dots are as up-close-and-personal as I wish to get to the human wreckage of a plane crash.

In the case of TWA Flight 800, Shanahan was on the trail of a bomb. He was analyzing the victims' injuries for evidence of an explosion in the cabin. If he found it, he would then try to pinpoint where on the plane the bomb had been. He takes a thick folder from a file cabinet drawer and pulls out his team's report. Here is the chaos and gore of a major passenger airline crash quantified and outlined, with figures and charts and bar graphs, transformed from horror into something that can be discussed over coffee in a National Transportation Safety Board morning meeting. "4.19: Injury Predominance Right vs. Left with Floating Victims." "4.28: Mid-Shaft Femur Fractures and Forward Horizontal Seat Frame Damage." I ask Shanahan whether the statistics and the dispassionate prose helped him maintain what I imagine to be a necessary emotional remove from the human tragedy behind the inquiry. He looks down at his hands, which rest, fingers interlinked, on the Flight 800 folder.

"Maureen will tell you I coped variably with Flight 800. It was emotionally very traumatic, particularly with the number of teenagers on board. A high school French club going to Paris. Young couples. We were all pretty grim." Shanahan says this isn't typical of the mood behind the scenes at a crash site. "You want a very superficial involvement, so jokes and light-heartedness tend to be fairly common. Not this time."

For Shanahan, the hardest thing about Flight 800 was that most of the bodies were relatively whole. "Intactness bothers me much more than the lack of it," he says. The sorts of things most of us can't imagine seeing or coping with—severed hands, legs, scraps of flesh—Shanahan is more comfortable with. "That way, it's just tissue. You can put yourself in that frame of mind and get on with your job." It's gory, but not sad. Gore you get used to. Shattered lives you don't. Shanahan does what the pathologists do. "They focus on the parts, not the person. During the autopsy, they'll be describing the eyes, then the mouth. You don't stand back and say, 'This is a person who is the father of four.' It's the only way you can emotionally survive."

Ironically, intactness is one of the most useful clues in determining whether a bomb has gone off. We are on page 16 of the report, Heading 4.7: Body Fragmentation. "People very close to an explosion come apart," Shanahan says to me quietly. Dennis has a way of talking about these things that seems neither patronizingly euphemistic nor offensively graphic. Had there been a bomb in the cabin of Flight 800, Shanahan would have found a cluster of "highly fragmented bodies" corresponding to the seats nearest the explosion. In fact, most of the bodies were primarily intact, a fact quickly gleaned by noting their body fragmentation code. To simplify the work of people like Shanahan who must analyze large numbers of reports, medical examiners often use color codes. On Flight 800, for instance, people ended up either Green (body intact), Yellow (crushed head or the loss of one extremity), Blue (loss of 2 extremities with or without crushed head), or Red (loss of 3 or more extremities or complete transection of body).

Another way the dead can help determine whether a bomb went off is through the numbers and trajectories of the "foreign bodies" embedded within them. These show up on X-rays, which are routinely taken as part of each crash autopsy. Bombs launch shards of themselves and of nearby objects into people seated close by; the patterns within each body and among the bodies overall can shed light on whether a bomb went off and where. If a bomb went off in a starboard bathroom, for instance, the people whose seats faced it would carry fragments that entered the fronts of their bodies. People across the aisle from it would display these injuries on their right sides. As Shanahan had expected, no telltale patterns emerged.

Shanahan turned next to the chemical burns found on some of the bodies. These burns had begun to fuel speculation that a missile had torn through the cabin. It's true that chemical burns in a crash are usually caused by contact with highly caustic fuel, but Shanahan suspected that the burns had happened after the plane hit the water. Spilled jet fuel on the surface of the water will burn a floating body on its back, but not on its front. Shanahan checked to be sure that all the "floaters"—people recovered from the water's surface—were the ones with the chemical burns, and that these burns were on their backs. And they were. Had a missile blasted through the cabin, the fuel burns would have been on people's fronts or sides, depending on where they had been seated, but not their backs, as the seatbacks would have protected them. No evidence of a missile.

Shanahan also looked at thermal burns, the kind caused by fire. Here there was a pattern. By looking at the orientation of the burns—most were on the front of the body—he was able to trace the path of a fire that had swept through the cabin. Next he looked at data on how badly these passengers' seats had been burned. That their chairs were far more severely burned than they themselves were told him that people had been thrown from their seats and clear of the plane within seconds after the fire broke out. Authorities had begun to suspect that a wing fuel tank had exploded. The blast was far enough away from passengers that they had remained intact, but serious enough to damage the body of the plane to the point that it broke apart and the passengers were thrown clear.

I ask Shanahan why the bodies would be thrown from the plane if they were wearing seat belts. Once a plane starts breaking up, he replies, enormous forces come into play. Unlike the split-second forces of a bomb, they won't typically rip a body apart, but they are powerful enough to wrench passengers from their seats. "This is a plane that's traveling at three hundred miles per hour," Shanahan says. "When it breaks up, it

loses its aerodynamic capability. The engines are still providing thrust, but now the plane's not stable. It's going to be going
through horrible gyrations. Fractures propagate and within
five or six seconds this plane's in chunks. My theory is that the
plane was breaking up pretty rapidly, and seatbacks were collapsing and people were slipping out of their restraint systems."

The Flight 800 injuries fit Dennis's theory: People tended
to have the sort of massive internal trauma that one typically
sees from what they call in Shanahan's world "extreme water
impact." A falling human stops short when it hits the surface
of the water, but its organs keep traveling for a fraction of a
second longer, until they hit the wall of the body cavity, which
by that point has started to rebound. The aorta often ruptures
because part of it is fixed to the body cavity—and thus stops
at the same time—while the other part, the part closest to the
heart, hangs free and stops slightly later; the two parts wind up
traveling in opposite directions and the resultant shear forces
cause the vessel to snap. Seventy-three percent of Flight 800's
passengers had serious aortic tears.

The other thing that reliably happens when a body hits
water after a long fall is that the ribs break. This fact has been
documented by former Civil Aeromedical Institute researchers Richard Snyder and Clyde Snow. In 1968, Snyder looked
at autopsy reports from 169 people who had jumped off the
Golden Gate Bridge. Eighty-five percent had broken ribs,
whereas only 15 percent emerged with fractured vertebrae and
only a third with arm or leg fractures. Broken ribs are minor
in and of themselves, but during high-velocity impacts they
become sharp, jagged weapons that pierce and slice what lies
within them: heart, lungs, aorta. In 76 percent of the cases
Snyder and Snow looked at, the ribs had punctured the lungs.
Statistics from Flight 800 sketched a similar scenario: Most of
the bodies displayed the telltale internal injuries of extreme

water impact. All had blunt chest injuries, 99 percent had multiple broken ribs, 88 percent had lacerated lungs, and 73 percent had injured aortas.

If a brutal impact against the water's surface was what killed most passengers, does that mean they were alive and aware of their circumstances during the three-minute drop to the sea? Alive, perhaps. "If you define alive as heart pumping and them breathing," says Shanahan, "there might have been a significant number." Aware? Dennis doesn't think so. "I think it's very remote. The seats and the passengers are being tossed around. You'd just get overwhelmed." Shanahan has made a point of asking the hundreds of plane and car crash survivors he interviews what they felt and observed during their accident. "I've come to the general conclusion that they don't have a whole lot of awareness that they've been severely traumatized. I find them very detached. They're aware of a lot of things going on, but they give you this kind of ethereal response—'I knew what was going on, but I didn't really know what was going on. I didn't particularly feel like I was a part of it, but on the other hand I knew I was a part of it.'"

Given that so many Flight 800 passengers were thrown clear of the plane as it broke apart, I wondered whether they stood a chance—however slim—of surviving. If you hit the water like an Olympic diver, might it be possible to survive a fall from a high-flying plane? It has happened at least once. In 1963, our man of the long-distance plummet, Richard Snyder, turned his attention to people who had survived falls from normally fatal heights. In "Human Survivability of Extreme Impacts in Free-Fall," he reports the case of a man who fell seven miles from an airplane and survived, albeit for only half a day. And this poor sap didn't have the relative luxury of a water landing. He hit ground. (From that height, in fact, there is little difference.) What Snyder found is that a person's speed at impact doesn't dependably predict the severity of his or her injuries.

He spoke with eloping bridegrooms who sustained more debilitating injuries falling off their ladders than did a suicidal thirty-six-year-old who dropped seventy-one feet onto concrete. The latter walked away needing nothing more than Band-Aids and a therapist.

Generally speaking, people falling from planes have booked their final flight. According to Snyder's paper, the maximum speed at which a human being has a respectable shot at surviving a feet-first—that's the safest position—fall into water is about 70 mph. Given that the terminal velocity of a falling body is 120 mph, and that it takes only five hundred feet to reach that speed, you are probably not going to fall five miles from an exploding plane and live to be interviewed by Dennis Shanahan.

Was Shanahan right about Flight 800? He was. Over time, critical pieces of the plane were recovered, and the wreckage supported his findings. The final determination: Sparks from frayed wiring had ignited fuel vapors, causing an explosion of one of the fuel tanks.

The unjolly science of injury analysis got its start in 1954, the year two British Comet airliners mysteriously dropped from the sky into the sea. The first plane vanished in January, over Elba, the second off Naples three months later. In both crashes, owing to the depth of the water, authorities were unable to recover much of the wreckage and so turned for clues to the "medical evidence": the injuries of the twenty-one passengers recovered from the surface of the sea.

The investigation was carried out at Britain's Royal Air Force Institute of Aviation Medicine in Farnborough, by the organization's group captain, W. K. Stewart, in conjunction with one Sir Harold E. Whittingham, director of medical services for the British Overseas Airways Corporation. As Sir

Harold held the most degrees—five are listed on the published paper, not counting the knighthood—I will, out of respect, assume him to have been the team leader.

Sir Harold and his team were immediately struck by the uniformity of the corpses' injuries. All twenty-one cadavers showed relatively few external wounds and quite severe internal injuries, particularly to the lungs. Three conditions were known to cause lung injuries such as those found in the Comet bodies: bomb blast, sudden decompression—as happens when pressurization of an airplane cabin fails—and a fall from extreme heights. Any one of them, in a crash like these, was a possibility. So far, the dead weren't doing much to clear up the mystery.

The bomb possibility was the first to be ruled out. None of the bodies were burned, none had been penetrated with bomb-generated shrapnel, and none had been, as Dennis Shanahan would put it, highly fragmented. The insane, grudge-bearing, explosives-savvy former Comet employee theory quickly bit the dust.

Next the team considered sudden depressurization of the passenger cabin. Could this possibly cause such severe lung damage? To find out, the Farnborough team recruited a group of guinea pigs and exposed them to a sudden simulated pressure drop—from sea level to 35,000 feet. To quote Sir Harold, "The guinea pigs appeared mildly startled by the experience but showed no signs of respiratory distress." Data from other facilities, based on both animal experimentation and human experiences, showed similarly few deleterious effects—certainly not the kind of damage seen in the lungs of the Comet passengers.

This left our friend "extreme water impact" as the likely cause of death, and a high-altitude cabin breakup, presumably from some structural flaw, as the likely cause of the crash.

As Richard Snyder wouldn't write "Fatal Injuries Resulting from Extreme Water Impact" for another fourteen years, the Farnborough team turned once again to guinea pigs. Sir Harold wanted to find out exactly what happens to lungs that hit water at terminal velocity. When I first saw mention of the animals, I pictured Sir Harold trekking to the cliffs of Dover, rodent cages in tow, and hurling the unsuspecting creatures into the seas below, where his companions awaited in rowboats with nets. But Sir Harold had more sense than I; he and his men devised a "vertical catapult" to achieve the necessary forces in a far shorter distance. "The guinea pigs," he wrote, "were lightly secured by strips of adhesive paper to the under surface of the carrier so that, when the latter was arrested to the lower limit of its excursion, the guinea pig was projected belly first, about 2½ feet through the air before hitting the water." I know just the sort of little boy Sir Harold was.

To make a long story short, the catapulted guinea pigs' lungs looked a lot like the Comet passengers' lungs. The researchers concluded that the planes had broken apart at altitude, spilling most of their human contents into the sea. To figure out exactly where the fuselage had broken apart, they looked at whether the passengers had been clothed or naked when pulled from the sea. Sir Harold's theory was that hitting the sea from a height of several miles would knock one's clothes off, but that hitting the sea inside the largely intact tail of the plane would not, and that they could therefore surmise the point of breakup as the dividing line between clothed and naked cadavers. For in both flights, it was the passengers determined (by checking the seating chart) to have been in the back of the plane who wound up floating in their clothes, while passengers seated forward of a certain point were found floating naked, or practically so.

To prove his theory, Sir Harold lacked one key piece of data:

Was it indeed true that hitting the sea after falling from an air-plane would serve to knock one's clothes off? Ever the pioneer, Sir Harold undertook the study himself. Though I would like nothing better than to be able to relate to you the details of another Farnborough guinea pig study, this one featuring the little rodents outfitted in tiny worsted suits and 1950s dresses, in point of fact no guinea pigs were used. The Royal Aircraft Establishment was enlisted to pilot a group of fully clothed dummies to cruising altitude and drop them into the sea.* As Sir Harold had expected, their clothes were indeed blown off on impact, a phenomenon verified by Marin County coroner Gary Erickson, the man who autopsies the bodies of Golden Gate Bridge suicides: Even after falling just 250 feet, he told me, "typically the shoes get blown off, the crotch gets blown out of the pants, one or both of the rear pockets are gone."

In the end, enough of the Comet wreckage was recovered to verify Sir Harold's theories. A structural failure had indeed caused both planes to break apart in midair. Hats off to Sir Harold and the guinea pigs of Farnborough.

Dennis and I are eating an early lunch at an Italian restaurant near the beach. We are the only customers, and it's way too quiet for the conversation going on at our table. Whenever the waiter appears to refill our water glasses, I pause, as though we were discussing something top secret or desperately personal. Shana-han seems not to care. The waiter will be grinding pepper on my

---

* You are perhaps wondering, as I did, whether cadavers were ever used to doc-ument the effects of accidental free falls on humans. The closest I came to a paper like this was J. C. Earley's "Body Terminal Velocity," dated 1964, and J. S. Cot-ner's "Analysis of Air Resistance Effects on the Velocity of Falling Human Bod-ies," from 1962, both, alas, unpublished. I do know that when J. C. Earley used dummies in a study, he used "Dummies" in the title, and so I suspect that a few donated corpses did indeed make the plunge for science.

salad for what seems like a week, and Dennis is going, ". . . used a scallop trawler to recover some of the smaller remains . . ."

I ask Dennis how, knowing what he knows and seeing what he sees, he ever manages to board a plane. He points out that most crashing airplanes don't hit the ground from thirty thousand feet. The vast majority crash on takeoff or landing, either on or near the ground. Shanahan says 80 to 85 percent of plane crashes are potentially survivable.

The key word here is "potentially." Meaning that if everything goes the way it went in the FAA-required cabin evacuation simulation, you'll survive. Federal regulations require airplane manufacturers to be able to evacuate all passengers through half of a plane's emergency exits within ninety seconds. Alas, in reality, evacuations rarely happen the way they do in simulations. "If you look at survivable crashes, it's rare that even half the emergency exits open," says Shanahan. "Plus, there's a lot of panic and confusion." Shanahan cites the example of a Delta crash in Dallas. "It should have been very survivable. There were very few traumatic injuries. But a lot of people were killed by the fire. They found them stacked up at the emergency exits. Couldn't get them open." Fire is the number one killer in airplane mishaps. It doesn't take much of an impact to explode a fuel tank and set a plane on fire. Passengers die from inhaling searing-hot air and from toxic fumes released by burning upholstery or insulation. They die because their legs are broken from slamming into the seat in front of them and they can't crawl to the exits. They die because passengers don't exit flaming planes in an orderly manner; they stampede and elbow and trample.*

---

* Here is the secret to surviving one of these crashes: Be male. In a 1970 Civil Aeromedical Institute study of three crashes involving emergency evacuations, the most prominent factor influencing survival was gender (followed closely by proximity to exit). Adult males were by far the most likely to get out alive. Why? Presumably because they pushed everyone else out of the way.

Could airlines do a better job of making their planes fire-safe? You bet they could. They could install more emergency exits, but they won't, because that means taking out seats and losing revenue. They could install sprinkler systems or build crash-worthy fuel systems of the type used on military helicopters. But they won't, because both these options would add too much weight. More weight means higher fuel costs.

Who decides when it's okay to sacrifice human lives to save money? Ostensibly, the Federal Aviation Administration. The problem is that most airline safety improvements are assessed from a cost-benefit viewpoint. To quantify the "benefit" side of the equation, a dollar amount is assigned to each saved human life. As calculated by the Urban Institute in 1991, you are worth $2.7 million. "That's the economic value of the cost of somebody dying and the effects it has on society," said Van Goudy, the FAA man I spoke with. While this is considerably more than the resale value of the raw materials, the figure in the benefits column is rarely large enough to surpass the airlines' projected costs. Goudy used the example of shoulder harnesses, which I had asked him about. "The agency would say, 'All right, if you're going to save fifteen lives over the next twenty years by putting in shoulder straps, that's fifteen times two million dollars; that's thirty million.' The industry comes back and says, 'It's gonna cost us six hundred and sixty-nine million to put the things in.'" So long, shoulder straps.

Why doesn't the FAA then come back and say, "Tough tiddlywinks. You're putting them in anyway"? For the same reason it took fifteen years for the government to begin requiring air bags in cars. The regulatory agencies have no teeth. "If the FAA wants to promulgate a regulation, they have to provide the industry with a cost-benefit analysis and send it out for comment," says Shanahan. "If the industry doesn't like what they see, they go to their congressmen. If you're Boeing, you have a tremendous influence in Congress."*

To the FAA's credit, the agency recently approved a new "inerting" system that pumps nitrogen-enriched air into fuel tanks, reducing the levels of highly flammable oxygen and the likelihood of an explosion such as the one that brought down Flight 800.

I ask Dennis whether he has any advice for the people who'll read this book and never again board a plane without wondering if they're going to wind up in a heap of bodies at the emergency exit door. He says it's mostly common sense. Sit near an emergency exit. Get down low, below the heat and smoke. Hold your breath as long as you can, so you don't cook your lungs and inhale poisonous fumes. Shanahan prefers window seats because people seated on the aisle are more likely to get beaned with the suitcases that can come crashing through the overhead bin doors in even a fairly mild impact.

As we wait for the bill, I ask Shanahan the question he gets asked at every cocktail party he's been to in the past twenty years: Are your chances of surviving a crash better near the front of the plane or the back? "That depends," he says patiently, "on what kind of crash it's going to be." I rephrase the question. Given his choice of anywhere on the plane, where does he prefer to sit?

"First class."

---

* This is no doubt why planes today are not equipped with air bags. Believe it or not, someone actually designed an airplane air-bag system, called the Airstop Restraint System, which combined underfoot, underseat, and chest air bags. The FAA even tested the system on dummies on a DC-7 that it crashed into a hill outside of Phoenix, Arizona, in 1964. While a control dummy in a lap belt fastened low and tight about it jackknifed violently and lost its head, the Airstop-protected dummy fared just fine. The designers were inspired by stories of World War II fighter pilots who would inflate their life vests just before a crash.

6

THE CADAVER WHO
JOINED THE ARMY

The sticky ethics of bullets and bombs

For three days in January of 1893 and again for four days in March, Captain Louis La Garde of the U.S. Army Medical Corps took up arms against a group of extraordinary foes. It was an unprecedented military undertaking, and one for which he would forever after be remembered. Though La Garde served as a surgeon, he was no stranger to armed combat. In the Powder River Expedition of 1876, he had been decorated for gallantry in confronting tribes of hostile Sioux. La Garde had led the charge against Chief Dull Knife, whose name, we can only assume, was no reflection on his intellectual and military acumen or the quality and upkeep of his armaments.

La Garde received his strange and fateful orders in July of 1892. He would be receiving, the letter said, a new, experimental .30-caliber Springfield rifle. He was to take this rifle, along with his standard-issue .45-caliber Springfield, and report to Frankford Arsenal, Pennsylvania, the following winter. Taking shape in the rifles' sights would be men, a series of them, naked and unarmed. That they were naked and unarmed was the less distinctive thing about them. More distinctive was that they were already dead. They had died of natural causes and had been collected—from where is not revealed—as subjects in an Army Ordnance Department experiment. They were to be suspended from a tackle in the ceiling of the firing range, shot at in a dozen places and with a dozen different charges (to simulate different distances), and autopsied. La Garde's mission was to compare the physiological effects of the two different weapons upon the human body's bones and innards.

The United States Army was by no means the first to sanction the experimental plugging of civilian cadavers. The French army, wrote La Garde in his book *Gunshot Injuries,* had been "firing into dead bodies for the purpose of teaching the effects of gunshots in war" since around 1800. Ditto the Germans, who went to the exquisite trouble of propping up their mock victims al fresco, at distances approximating those of an actual battlefield. Even the famously neutral Swiss sanctioned a series of military wound ballistics studies on cadavers in the late 1800s. Theodore Kocher, a Swiss professor of surgery and a member of the Swiss army militia (the Swiss prefer not to fight, but they are armed, and with more than little red pocket knife/can openers), spent a year firing Swiss Vetterli rifles into all manner of targets—bottles, books, water-filled pig intestines, oxen bones, human skulls, and, ultimately, a pair of whole human cadavers—with the aim of understanding the mechanisms of wounding from bullets.

Kocher—and to a certain extent La Garde—expressed a desire that their ballistics work with cadavers would lead to a more humanitarian form of gun battle. Kocher urged that the goal of warfare be to render the enemy not dead, but simply unable to fight. To this end, he advised limiting the size of the bullets and making them from a material of a higher melting point than lead, so that they would deform less and thus destroy less tissue.

Incapacitation—or stopping power, as it is known in munitions circles—became the Holy Grail of ballistics research. How to stop a man in his tracks, preferably without maiming or killing him, but definitely before he maimed or killed you first. Indeed, the next time Captain La Garde and his swinging cadavers took the stage, in 1904, it was in the name of improved stopping power. The topic had been high on the generals' to-do lists following the army's involvement in the Philippines, during the final stage of the Spanish-American War, where its Colt .38s had failed, on numerous occasions, to stop the enemy cold. While the Colt .38 was considered sufficient for "civilized" warfare—"even the stoical Japanese soldier," wrote La Garde in *Gunshot Injuries*, "fell back as a rule when he was hit the first time"—such was apparently not the case with "savage tribes or a fanatical enemy." The Moro tribesman of the Philippines was considered a bit of both: "A fanatic like a Moro, wielding a bolo in each hand who advances with leaps and bounds . . . must be hit with a projectile having a maximum amount of stopping power," wrote La Garde. He related the tale of one battle-enlivened tribesman who charged a U.S. Army guard unit. "When he was within 100 yards, the entire guard opened fire on him." Nonetheless, he managed to advance some ninety-five yards toward them before finally crashing to the ground.

La Garde, at the War Department's urging, undertook an investigation of the army's various guns and bullets and their relative efficacy at putting a rapid halt to enemies. He decided that one way to do this would be to fire at suspended cadavers and take note of the "shock," as estimated by "the disturbance which appeared." In other words, how far back does the hanging torso or arm or leg swing when you shoot it? "It was based on the assumption that the momentum of hanging bodies of various weights could somehow be correlated and measured, and that it actually meant something with regard to stopping power," says Evan Marshall, who wrote the book on handgun stopping power (it's called *Handgun Stopping Power*). "What it actually did was extrapolate questionable data from questionable tests."

Even Captain La Garde came to realize that if you want to find out how likely a gun is to stop someone, you are best off trying it on an entity that isn't already quite permanently stopped. In other words, a live entity. "The animals selected were beeves about to undergo slaughter in the Chicago stockyards," wrote La Garde, deeply perplexing the ten or fifteen people who would be reading his book later than the 1930s, when the word "beeves," meaning cattle, dropped from everyday discourse. Sixteen beeves later, La Garde had his answer: Whereas the larger-caliber (.45) Colt revolver bullets caused the cattle to drop to the ground after three or four shots, the animals shot with smaller-caliber .38 bullets failed even after ten shots to drop to the ground. And ever since, the U.S. Army has gone confidently into battle, knowing that when cows attack, their men will be ready.

For the most part, it has been the lowly swine that has borne the brunt of munitions trauma research in the United States and Europe. In China—at the No. 3 Military Medical College and the China Ordnance Society, among others—it has been mongrel dogs that get shot at. In Australia, as reported in the Proceedings of the 5th Symposium on Wound Ballistics, the researchers took aim at rabbits. It is tempting to surmise that a culture chooses its most reviled species for ballistics research. China occasionally eats its dogs, but doesn't otherwise have much use or affection for them; in Australia, rabbits are considered a scourge—imported by the British for hunting, they multiplied (like rabbits) and, in a span of twenty years, wiped out two million acres of south Australian brush.

In the case of the U.S. and European research, the theory doesn't hold. Pigs don't get shot at because our culture reviles them as filthy and disgusting. Pigs get shot at because their organs are a lot like ours. The heart of the pig is a particularly close match. Goats were another favorite, because their lungs are like ours. I was told this by Commander Marlene DeMaio, who studies body armor at the Armed Forces Institute of Pathology (AFIP). Talking to DeMaio, I got the impression that it would be possible to construct an entire functioning nonhuman human from pieces of other species. "The human knee most resembles the brown bear's," she said at one point, following up with a surprising or not so surprising statement: "The human brain most resembles that of Jersey cows at about six months."[*] I learned elsewhere that emu hips are dead ringers for human hips, a situation that has worked out better for humans than for emus, who, over at Iowa State University, have been lamed in a manner that mimics osteonecrosis and then shuttled in and out of CT scanners by researchers seeking to understand the disease.

Had I been calling the shots back at the War Department, I would have sanctioned a study not on why men sometimes fail to drop to the ground after being shot, but on why they so often *do*. If it takes ten or twelve seconds to lose consciousness from blood loss (and consequent oxygen deprivation to the brain), why, then, do people who have been shot so often collapse on the spot? It doesn't happen just on TV.

I posed this question to Duncan MacPherson, a respected ballistics expert and consultant to the Los Angeles Police Department. MacPherson insists the effect is purely psychological. Whether or not you collapse depends on your state of mind. Animals don't know what it means to be shot, and, accordingly, rarely exhibit the instant stop-and-drop. MacPherson points out that deer shot through the heart often run off for forty or fifty yards before collapsing. "The deer doesn't know anything about what's going on, so he just does

---

[*] I did not ask DeMaio about sheep and the purported similarity of portions of their reproductive anatomy to that of the human female, lest she be forced to draw conclusions about the similarity of my intellect and manners to that of the, I don't know, boll weevil.

his deer thing for ten seconds or so and then he can't do it anymore. An animal with a meaner disposition will use that ten seconds to come at you." On the flip side, there are people who are shot at but not hit—or hit with nonlethal bullets, which don't penetrate, but just smart a lot—who will drop immediately to the ground. "There was an officer I know who took a shot at a guy and the guy just went flat, totally splat, facedown," MacPherson told me. "He said to himself, 'God, I was aiming for center mass like I'm supposed to, but I must have gotten a head shot by mistake. I'd better go back to the shooting range.' Then he went to the guy and there wasn't a mark on him. If there isn't a central nervous system hit, anything that happens fast is all psychological."

MacPherson's theory would explain the difficulties the army had in La Garde's day with the Moro tribesmen, who presumably weren't familiar with the effects of rifles and kept on doing their Moro tribesman thing until they couldn't— owing to blood loss and consequent loss of consciousness— do it anymore. Sometimes it isn't just ignorance as to what bullets do that renders a foe temporarily impervious. It can also be viciousness and sheer determination. "A lot of guys take pride in their imperviousness to pain," MacPherson said. "They can get a lot of holes in them before they go down. I know an LAPD detective who got shot through the heart with a .357 Magnum and he killed the guy that shot him before he collapsed."

Not everyone agrees with the psychological theory. There are those who feel that some sort of neural overload takes place when a bullet hits. I communicated with a neurologist/avid handgunner/reserve deputy sheriff in Victoria, Texas, named Dennis Tobin, who has a theory about this. Tobin, who wrote the chapter "A Neurologist's View of 'Stopping Power'" in the book *Handgun Stopping Power*, posits that an area of the brain stem called the reticular activating system (RAS) is responsible for the sudden collapse. The RAS can be affected by impulses arising from massive pain sensations in the viscera.* Upon receiving these impulses, the RAS sends out a signal that weakens certain leg muscles, with the result that the person drops to the ground.

Somewhat shaky support for Tobin's neurological theory can be found in animal studies. Deer may keep going, but dogs and pigs seem to react as humans do. The phenomenon was remarked upon in military medical circles as far back as 1893. A wound ballistics experimenter by the name of Griffith, while going about his business documenting the effects of a Krag-Jorgensen rifle upon the viscera of live dogs at two hundred yards, noted that the animals, when shot in the abdomen, "died as promptly as though they had been electrocuted." Griffith found this odd, given that, as he pointed out in the *Transactions of the First Pan-American Medical Congress*, "no vital part was hit which might account for the instantaneous death of the animals." (In fact, the dogs were probably not as promptly dead as Griffith believed. More likely, they had simply collapsed and looked, from two hundred yards, like dead dogs. And by the time Griffith had walked the two hundred yards to get to them, they were in fact dead dogs, having expired from blood loss.)

---

* MacPherson counters that bullet wounds are rarely, at the outset, painful. Research by eighteenth-century scientist/philosopher Albrecht von Haller suggests that it depends on what the bullet hits. Experimenting on live dogs, cats, rabbits, and other small unfortunates, Haller systematically catalogued the viscera according to whether or not they register pain. By his reckoning, the stomach, intestines, bladder, ureter, vagina, womb, and heart do, whereas the lungs, liver, spleen, and kidneys "have very little sensation, seeing I have irritated them, thrust a knife into them, and cut them to pieces without the animals' seeming to feel any pain." Haller admitted that the work suffered certain methodological shortcomings, most notably that, as he put it, "an animal whose thorax is opened is in such violent torture that it is hard to distinguish the effect of an additional slight irritation."

In 1988, a Swedish neurophysiologist named A. M. Göransson, then of Lund University, took it upon himself to investigate the conundrum. Like Tobin, Göransson figured that
something about the bullet's impact was causing a massive
overload to the central nervous system. And so, perhaps
unaware of the similarities between the human brain and
that of Jersey cows at six months, he wired the brains of nine
anesthetized pigs to an EEG machine, one at a time, and shot
them in the hindquarters. Göransson reports having used a
"high-energy missile" for the task, which is less drastic than
it suggests. What it suggests is that Dr. Göransson got into his
car, drove some distance from his laboratory, and launched
the Swedish equivalent of Tomahawk missiles at the hapless
swine, but in fact, I am told, the term simply means a small,
fast-moving bullet.

Instantly upon being hit, all but three of the pigs showed
significantly flattened EEGs, the amplitude in some cases having dropped by as much as 50 percent. As the pigs had already
been stopped in their tracks by the anesthesia, it is impossible
to say whether they would have been rendered so by the shots,
and Göransson opted not to speculate. And if they had lost
consciousness, Göransson had no way of knowing what the
mechanism was. To the deep chagrin of pigs the world over,
he encouraged further study.

Proponents of the neural overload theory point to the "temporary stretch cavity" as the source of the effect. All bullets,
upon entry into the human form, blow open a cavity in the
tissue around them. This cavity shuts back up almost immediately, but in that fraction of a second that it is agape, the
nervous system, they believe, issues a Mayday blast—enough
of one, it seems, to overload the circuits and cause the whole
system to hang a Gone Fishing sign on the door.

These same proponents believe that bullets that create sizable stretch cavities are thus more likely to deliver the nec-

essary shock to achieve the vaunted ballistics goal of "good stopping power." If this is true, then in order to gauge a bullet's stopping power, one needs to be able to view the stretch cavity as it opens up. That is why the good Lord, working in tandem with the Kind & Knox gelatin company, invented human tissue simulant.

I am about to fire a bullet into the closest approximation of a human thigh outside of a human thigh: a six-by-six-by-eighteen-inch block of ballistic gelatin. Ballistic gelatin is essentially a tweaked version of Knox dessert gelatin. It is denser than dessert gelatin, having been formulated to match the average density of human tissue, is less colorful, and, lacking sugar, is even less likely to please dinner guests. Its advantage over a cadaver thigh is that it affords a stop-action view of the temporary stretch cavity. Unlike real tissue, human tissue simulant doesn't snap back: The cavity remains, allowing ballistics types to judge, and preserve a record of, a bullet's performance. Plus, you don't need to autopsy a block of human tissue simulant; because it's clear, you just walk up to it after you've shot it and take a look at the damage. Following which, you can take it home, eat it, and enjoy stronger, healthier nails in thirty days.

Like other gelatin products, ballistic gelatin is made from processed cow bone chips and "freshly chopped" pig hide.* The Kind & Knox Web site does not include human tissue simulant on its list of technical gelatin applications, which

---

* According to the Kind & Knox Web site, other products made with cow-bone-and-pigskin-based gelatin include marshmallows, nougat-type candy bar fillings, liquorice, Gummi Bears, caramels, sports drinks, butter, ice cream, vitamin gel caps, suppositories, and that distasteful whitish peel on the outside of salamis. What I am getting at here is that if you're going to worry about mad cow disease, you probably have more to worry about than you thought. And that if there's any danger, which I like to think there isn't, we're all doomed, so relax and have another Snickers.

rather surprised me, as did the failure of a Knox public relations woman to return my calls. You would think that a company that felt comfortable extolling the virtues of Number 1 Pigskin Grease on its Web site ("It is a very clean material"; "Available in tanker trucks or railcars") would be okay with talking about ballistic gelatin, but apparently I've got truckloads or railcars to learn about gelatin PR.

Our replicant human thigh was cooked up by Rick Lowden, a freewheeling materials engineer whose area of expertise is bullets. Lowden works at the Department of Energy's Oak Ridge National Laboratory in Oak Ridge, Tennessee. The lab is best known for its plutonium work in the Manhattan (atomic bomb development) Project and now covers a far broader and generally less unpopular range of projects. Lowden, for instance, has lately been involved in the design of an environmentally friendly no-lead bullet that doesn't cost the military an arm and a leg to clean up after. Lowden loves guns, loves to talk about them. Right now he's trying to talk about them with me, a distinctly trying experience, for I keep shepherding the conversation back to dead bodies, which Lowden clearly doesn't enjoy very much. You would think that a man who felt comfortable extolling the virtues of hollow-point bullets ("expands to twice its size and just thumps that person") would be okay talking about dead bodies, but apparently not. "You just cringe," he said, when I mentioned the prospect of shooting into human cadaver tissue. Then he made a noise that I transcribed in my notes as "Olggh."

We are standing under a canopy at the Oak Ridge shooting range, about to set up the first stopping-power test. The "thighs" sit in an open plastic cooler at our feet, sweating mildly. They are consommé-colored and, owing to the cinnamon added to mask the material's mild rendering-plant effluvium, smell like Big Red chewing gum. Rick carries the cooler out to the target table, thirty feet away, and settles an ersatz thigh into the gel cradle. I make conversation with Scottie Dowdell, who is supervising the shooting range today. He is telling me about the pine beetle epidemic in the area. I point to a stand of dead conifers in the woods a quarter mile back behind the target. "Like over there?" Scottie says no. He says they died of bullet wounds, something I never knew pine trees could do.

Rick returns and sets up the gun, which isn't really a gun but a "universal receiver," a tabletop gun housing that can be outfitted with barrels of different calibers. Once it's aimed, you pull a wire to release the bullet. We're testing a couple of new bullets that claim to be frangible, meaning they break apart on impact. The frangible bullet was designed to solve the "overpenetration," or ricochet, problem, i.e., bullets passing through victims, bouncing off walls, and harming bystanders or the police or soldiers who fired them. The side effect of the bullet's breaking apart on impact is that it tends to do this inside your body if you're hit. In other words, it tends to have really, really good stopping power. It basically functions like a tiny bomb inside the victim and is therefore, to date, mainly reserved for "special response" SWAT-type activities, such as hostage rescue.

Rick hands me the trigger string and counts down from three. The gelatin sits on the table, soaking up the sunshine, basking beneath the calm, blue Tennessee skies—*tra la la, life is gay, it's good to be a gelatin block, I* . . . BLAM!

The block flips up into the air, off the table, and onto the grass. As John Wayne said, or would have, had he had the opportunity, this block of gelatin won't be bothering anyone anytime soon. Rick picks up the block and places it back in its cradle. You can see the bullet's journey through the "thigh." Rather than overpenetrating and exiting the back side, the bullet has stopped short several inches into the block. Rick points to the stretch cavity. "Look at that. A total dump of energy. Total incapacitation."

I had asked Lowden whether munitions professionals ever concern themselves, as did Kocher and La Garde, with trying to design bullets that will incapacitate without maiming or killing. Lowden's face displayed the sort of look it displayed earlier when I'd said that armor-piercing bullets were "cute." He answered that the military chooses weapons more or less by how much damage they can inflict on a target, "whether the target be a human or a vehicle." This is another reason ballistic gelatin tends to get used in stopping-power tests, rather than cadavers. We're not talking about research that will help mankind save lives; we're talking about research that will help mankind take lives. I suppose you could argue that policemen's and soldiers' lives may be saved, but only by taking someone else's life first. Anyway, it's not a use of human tissue for which you're likely to get broad public support.

Of course, the other big reason munitions people shoot ballistic gelatin is reproducibility: Provided you follow the recipe, it's always the same. Cadaver thighs vary in density and thickness, according to the age, gender, and physical condition of their owners when they stopped using them. Still another reason: Cleanup's a breeze. The remains of this morning's thighs have been picked up and repacked in the cooler, a tidy, bloodless mass grave of low-calorie dessert.

Not that a ballistic gelatin shootout is completely devoid of gore. Lowden points to the toe of my sneaker, at a *Pulp Fiction* fleck of spatter. "You got some simulant on your shoe."

Rick Lowden never shot a dead man, though he had his chance. He was working on a project, in cooperation with the University of Tennessee's human decay facility, aimed at developing bullets that would resist corrosion from the acid breakdown products inside a dead body and help forensics types solve crimes long after they happen.

Rather than shooting the experimental bullets into his cadavers, Lowden got down on his hands and knees with a scalpel and a pair of tweezers and surgically placed them. He explained that he did this because he wanted the bullets to end up in specific places: muscle, fatty tissue, the head and chest cavities, the abdomen. If he'd shot them into the tissue, they might have overpenetrated, as they say, and wound up in the dirt.

He also did it that way because he felt he had to. "It was always my impression that we couldn't shoot a body." He recalls another project, one in which he was developing a simulated human bone that could be put inside blocks of ballistic gelatin, much as banana and pineapple chunks are floated inside Jell-O. To calibrate the simulated bone, he needed to shoot some actual bone and compare the two. "I was offered sixteen cadaver legs to shoot at. DOE told me they would terminate my project if I did that. We had to shoot pig femurs instead."

Lowden told me that military munitions professionals even worry about the politics of shooting into freshly killed livestock. "A lot of guys won't do that. They'll go get a ham from the store or a leg from the slaughterhouse. Even then, a lot of them don't openly publish what they do. There's still a stigma."

Ten feet behind us, sniffing the air, is a groundhog who has made unfortunate real estate choices in his life. The animal is half the size of a human thigh. If you shot that groundhog with one of these bullets, I say to Rick, what would happen? Would it completely vaporize? Rick and Scottie exchange a look. I get the feeling that the stigma attached to shooting groundhogs is fairly minimal.

Scottie shuts the ammo case. "Create a lot of paperwork, is what would happen."

Only recently has the military dipped its toes back into the roiling waters of publicly funded cadaveric ballistics research.

As one would imagine, the goals are strictly humanitarian. At the Armed Forces Institute of Pathology's Ballistic Missile Trauma Research Lab last year, Commander Marlene DeMaio dressed cadavers in a newly developed body armor vest and fired a range of modern-day munitions at their chests. The idea was to test the manufacturer's claims before outfitting the troops. Apparently body armor manufacturers' effectiveness claims aren't always to be trusted. According to Lester Roane, chief engineer at the independent ballistics and body armor test facility H. P. White Labs, the companies don't do cadaver tests. H. P. White doesn't either. "Anybody looking at it coldly and logically shouldn't have any problem with it," said Roane. "It's dead meat. But for some reason, it's just something that has been politically incorrect from before there was a term for politically correct."

DeMaio's cadaver tests represent a distinct improvement over how vests were originally tested by the military: In Operation Boar, during the Korean War, the Doron vest was tested simply by giving it to six thousand soldiers and seeing how they fared compared to soldiers wearing standard vests. Roane says he once watched a video made by a Central American police department that tested their vests by having officers put them on and then shooting at them.

The trick to designing body armor is to make it thick and unyielding enough to stop bullets without making it so heavy and hot and uncomfortable that officers won't wear it. What you don't want is what the Gilbertese Islanders used to have. While I was in D.C. to see DeMaio, I stopped at the Smithsonian's Museum of Natural History, where I saw a display of body armor from the Gilbert Islands. Battles in Micronesia were so pitched and bloody that Gilbertese warriors would outfit themselves head to foot with doormat-thick armor fashioned from the twisted fibers of coconut hulls. On top of the significant humiliation of making one's entrance onto the battlefield looking like an enormous macramé planter was the fact that the armor was so bulky it required the assistance of several squires to help maneuver you.

As with automotive cadavers, DeMaio's body-armor bodies were instrumented with accelerometers and load cells, in this case on the sternum, to record the impact forces and give researchers a detailed medical rendering of what was happening to the chest inside the armor. With some of the nastier-caliber weapons, the cadavers sustained lung lacerations and rib fractures, but nothing that translated into an injury that—if you weren't already a cadaver—could kill you. More tests are planned, with the goal of making a test dummy along the lines of those used by the automotive industry—so that one day cadavers won't be needed.

Because she had proposed to use human cadavers, DeMaio was advised to proceed with extreme caution. She gathered the blessings of three institutional review boards, a military legal counsel, and a card-carrying ethicist. The project was ultimately approved, with one stipulation: no penetration. The bullets had to stop short of the cadavers' skin.

Did DeMaio roll her eyes in exasperation? She says not. "When I was in medical school I used to think, 'Come on, don't be irrational. They've expired, they've donated their bodies, you know?' When I got into this project I realized that we are part of the public trust, and even if it doesn't make scientific sense, we have to be responsive to people's emotional concerns."

On an institutional level, the caution comes from fear of liability and of unpleasant media reports and withdrawal of funding. I spoke with Colonel John Baker, the legal counsel from one of the institutions that sponsored DeMaio's research. The head of this institution preferred that I refrain from naming it and instead refer to it as simply "a federal institution in Washington." He told me that over the past twenty-some years, democratic congressmen and budget-minded legislators have tried to close the place down, as have Jimmy Carter, Bill Clinton, and People for the Ethical Treatment of Animals. I got the feeling that my request for an interview had brought this man's day crashing down like so many pine trees behind a DOE shooting range.

"The concern is that some survivor will be so taken aback that they'll bring suit," said Colonel Baker from his desk at a federal institution in Washington. "And there is no body of law in this area, nothing you can look to other than good judgment." He pointed out that although cadavers don't have rights, their family members do. "I could imagine some sort of lawsuit that is based upon emotional distress. . . . You get some of those [cases] in a cemetery situation, where the proprietor has allowed the coffins to rot away and the corpses pop up." I replied that as long as you have informed consent—a signed agreement from the donor stating that he has willed his body to medical research—it would seem that the survivors wouldn't have much of a case.

The sticking point is the word "informed." It's fair to say that when people donate remains, either their own or those of a family member, they usually don't care to know the grisly details of what might be done with them. And that if you did tell them the details, they might change their minds and withdraw consent. Then again, if you're planning to shoot guns at them, it might be good to run that up the flagpole and get the a-okay. "Part of respecting persons is telling them the information that they might have an emotional response to," says Edmund Howe, editor of the *Journal of Clinical Ethics*, who reviewed Marlene DeMaio's research proposal. "Though one could go the other way and spare them that response and therefore ethically not commit that harm. But the downside to withholding information that might be significant to them is that it would violate their dignity to an extent." Howe suggests a third possibility, that of letting the families make the choice: Would they prefer to hear the specifics of what is being done with the donated body—specifics that may be upsetting—or would they prefer not to know?

It's a delicate balance that, in the end, comes down to wording. Observes Baker, "You don't really want to be telling somebody, 'Well, what we'll be doing is dissecting their eyeballs. We take them out and put them on a table and then we dissect them into finer and finer parts and then once we're finished we scrape all that stuff up and put it into a biohazard bag and try to keep it together so we can return whatever's left to you.' That sounds horrible." On the other hand, "medical research" is a tad vague. "Instead, you say, 'One of our principal concerns here at the university is ophthalmology. So we do a lot here with ophthalmological materials.'" If someone cares to think it through, it isn't hard to come to the conclusion that someone in a lab coat will, at the very least, be cutting your eyeball out of your head. But most people don't care to think it through. They focus on the end, rather than the means: Someone's vision may one day be saved.

Ballistics studies are especially problematic. How do you decide it's okay to cut off someone's grandfather's head and shoot it in the face? Even when the reason you are doing that is to gather data to ensure that innocent civilians who are hit in the face with nonlethal bullets won't suffer disfiguring fractures? Moreover, how do you bring yourself to carry out the cutting off and shooting of someone's grandfather's head?

I posed these questions to Cindy Bir, who brought herself to do exactly that, and whom I met while I was at Wayne State. Bir is accustomed to firing projectiles at the dead. In 1993, the National Institute of Justice (NIJ) commissioned her to document the impact effects of various nonlethal munitions: plastic bullets, rubber ones, beanbags, the lot. Police began using non-lethal bullets in the late 1980s, in situations where they need to subdue civilians—mostly rioters and violent psychotics—without putting their lives in danger. In nine instances since that time, "nonlethal" bullets have proved lethal, prompting the NIJ to have Bir look into what was going on with these different bullets, with the aim of its not going on ever again.

As to the question "How do you bring yourself to cut off someone's grandfather's head?" Bir replied, "Thankfully, Ruhan does that for us." (The very same Ruhan who preps the cadavers for automotive impacts.) She added that the nonlethal munitions were not shot from guns but fired from air cannons, because doing so is both more precise and less disturbing. "Still," concedes Bir. "I was glad when that one finished up."

Bir copes like most other cadaver researchers do, with a mix of compassion and emotional remove. "You treat them with dignity, and you kind of separate the fact that . . . I don't want to say that they're not a person, but . . . you think of them as a specimen." Bir was trained as a nurse, and in some ways finds the dead easier to work with. "I know they can't feel it, and I know that I'm not going to hurt them." Even the most practiced cadaver researcher has days when the task at hand presents itself as something other than scientific method. For Bir, it had little to do with the fact that she was directing bullets at her subjects. It is the moments when the specimen steps out of his anonymity, his objecthood, and into his past existence as a human being.

"We received a specimen and I went down to help Ruhan, and this gentleman must have come directly from the nursing home or hospital," she recalls. "He had on a T-shirt and flannel PJ pants. It hit me like . . . this could be my dad. Then there was one that I went to look at—a lot of times you like to take a look at the specimen to make sure it's not too big [to lift]—and this person was wearing a hospital gown from my hometown."

If you really want to stay up late worrying about lawsuits and bad publicity, explode a bomb near the body of someone who has willed his remains to science. This is perhaps the most firmly entrenched taboo of the cadaveric research world. Indeed, live, anesthetized animals have generally been considered preferable, as targets of explosions, to dead human beings. In a 1968 Defense Atomic Support Agency paper entitled *Estimates of Man's Tolerance to the Direct Effects of Air Blast*—i.e., from bombs—researchers discussed the effects of experimental explosions upon mice, hamsters, rats, guinea pigs, rabbits, cats, dogs, goats, sheep, steers, pigs, burros, and stump-tailed macaques, but not upon the actual subject of inquiry. No one had ever strapped a cadaver up against the shock tube to see what might happen.

I called up a man named Aris Makris, who works for a company in Canada called Med-Eng Systems, which engineers protective gear for people who clear land mines. I told him about the DASA paper. Dr. Makris explained that dead people weren't always the best models for gauging living people's tolerance to explosive blasts because of their lungs, which are deflated and not doing the things that lungs normally do. The shock wave from a bomb wreaks the most havoc on the body's most easily compressed tissue, and that is found in the lungs: specifically, the tiny, delicate air sacs where the blood picks up oxygen and drops off carbon dioxide. An explosive shock wave compresses and ruptures these sacs. Blood then seeps into the lungs and drowns their owner, sometimes quickly, in ten or twenty minutes, sometimes over a span of hours.

Makris conceded that, biomedical issues aside, the blast tolerance chaps were probably not highly motivated to work with cadavers. "There are enormous ethical or PR challenges with that," he said. "It just hasn't been the habit of blasting cadavers: Please give your body to science so we can blow it up?"

One group recently braved the storm. Lieutenant Colonel Robert Harris and a team of other doctors from the Extremity Trauma Study Branch of the U.S. Army Institute of Surgical Research at Fort Sam Houston, Texas, recruited cadavers to test five types of footwear either commonly used by or being newly marketed for land mine clearance teams. Ever since the Vietnam War, a rumor had persisted that sandals were the safest footwear for land mine clearance, for they minimized injuries caused by fragments of the footwear itself being driven into the foot like shrapnel, compounding the damage and the risk of infection. Yet no one had ever tested the sandal claim on a real foot, nor had anyone done cadaver tests of any of the equipment being touted by manufacturers as offering greater safety than the standard combat boot.

Enter the fearless men of the Lower Extremity Assessment Program. Starting in 1999, twenty cadavers from a Dallas medical school willed body program were strapped, one by one, into a harness hanging from the ceiling of a portable blast shelter. Each cadaver was outfitted with strain gauges and load cells in its heel and ankle, and clad in one of six types of footwear. Some boots claimed to protect by raising the foot up away from the blast, whose forces attenuate quickly; others claimed to protect by absorbing or deflecting the blast's energy. The bodies were posed in standard walking position, heel to the ground, as though striding confidently to their doom. As an added note of verisimilitude, each cadaver was outfitted head to toe in a regulation battle dress uniform. In addition to the added realism, the uniforms conferred a measure of respect, the sort of respect that a powder-blue leotard might not, in the eyes of the U.S. Army anyway, supply.

Harris felt confident that the study's humanitarian benefits outweighed any potential breach of dignity. Nonetheless, he consulted the willed body program administrators about the possibility of informing family members about the specifics of the test. They advised against it, both because of what they called the "revisiting of grief" among families who had made piece with the decision to donate and because, when you get down to the nitty-gritty details of an experiment, virtually any use of a cadaver is potentially upsetting. If willed body program coordinators contacted the families of LEAP cadavers, would they then have to contact the families of the leg-drop-test cadavers down the hall, or, for that matter, the anatomy lab cadavers across campus? As Harris points out, the difference between a blast test and an anatomy class dissection is essentially the time span. One lasts a fraction of a second; the other lasts a year. "In the end," he says, "they look pretty much the same." I asked Harris if he plans to donate his body to research. He sounded downright keen on the prospect. "I'm always saying, 'After I die, just put me out there and blow me up.'"

If Harris could have done his research using surrogate "dummy" legs instead of cadavers, he would have done so. Today there are a couple good ones in the works, developed by the Australian Defence Science & Technology Organisation. (In Australia, as in other Commonwealth nations, ballistics and blast testing on human cadavers is not allowed. And certain words are spelled funny.) The Frangible Surrogate Leg (FSL) is made of materials that react to blast similarly to the way human leg materials do; it has mineralized plastic for bones, for example, and ballistic gelatin for muscle. In March of 2001, Harris exposed the Australian leg to the same land mine blasts that his cadavers had weathered, to see if the results correlated. Disappointingly, the bone fracture patterns were somewhat off. The main problem, at the moment, is cost. Each

FSL—they aren't reusable—costs around $5,000; the cost of a cadaver (to cover shipping, HIV and hepatitis C testing, cremation, etc.) is typically under $500.

Harris imagines it's only a matter of time before the kinks are worked out and the price comes down. He looks forward to that time. Surrogates are preferable not only because tests involving land mines and cadavers are ethically (and probably literally) sticky, but because cadavers aren't uniform. The older they are, the thinner their bones and the less elastic their tissue. In the case of land mine work, the ages are an especially poor match, with the average land mine clearer in his twenties and the average donated cadaver in its sixties. It's like market-testing Kid Rock singles on a roomful of Perry Como fans.

Until that time, it'll be rough going for Commonwealth land mine types, who cannot use whole cadavers. Researchers in the UK have resorted to testing boots on amputated legs, a much-criticized practice, owing to the fact that these limbs have typically had gangrene or diabetic complications that render them poor mimics of healthy limbs. Another group tried putting a new type of protective boot onto the hind leg of a mule deer for testing. Given that deer lack toes and heels and people lack hooves, and that no country I know of employs mule deer in land mine clearance, it is hard—though mildly entertaining—to try to imagine what the value of such a study could have been.

LEAP, for its part, turned out to be a valuable study. The sandal myth was mildly vindicated (the injuries were about as severe as they were with a combat boot), and one boot—Med-Eng's Spider Boot—showed itself to be a solid improvement over standard-issue footwear (though a larger sample is needed to be sure). Harris considers the project a success, because with land mines, even a small gain in protection can mean a huge difference in a victim's medical outcome. "If I can save a foot or keep an amputation below the knee," he says, "that's a win."

It is an unfortunate given of human trauma research that the things most likely to accidentally maim or kill people—things we most need to study and understand—are also the things most likely to mutilate research cadavers: car crashes, gunshots, explosions, sporting accidents. There is no need to use cadavers to study stapler injuries or human tolerance to ill-fitting footwear. "In order to be able to protect against a threat, whether it is automotive or a bomb," observes Makris, "you have to put the human to its limits. You've got to get destructive."

I agree with Dr. Makris. Does that mean I would let someone blow up my dead foot to help save the feet of NATO land mine clearers? It does. And would I let someone shoot my dead face with a nonlethal projectile to help prevent accidental fatalities? I suppose I would. What *wouldn't* I let someone do to my remains? I can think of only one experiment I know of that, were I a cadaver, I wouldn't want anything to do with. This particular experiment wasn't done in the name of science or education or safer cars or better-protected soldiers. It was done in the name of religion.

7

HOLY CADAVER

————

The crucifixion experiments

The year was 1931. French doctors and medical students were gathered in Paris for an annual affair called the Laennec conference. Late one morning, a priest appeared on the fringes of the gathering. He wore the long black cassock and Roman collar of the Catholic Church, and he carried a worn leather portfolio beneath one arm. His name was Father Armailhac, he said, and he sought the counsel of France's finest anatomists. Inside the portfolio was a series of close-up photographs of the Shroud of Turin, the linen cloth in which, believers held, Jesus had been wrapped for burial when he was taken down from the cross. The shroud's authenticity was in question then, as now, and the church had turned to medicine to see if the markings corresponded to the realities of anatomy and physiology.

Dr. Pierre Barbet, a prominent and none-too-humble surgeon, invited Father Armailhac to his office at Hôpital Saint-Joseph and swiftly nominated himself for the job. "I am . . . well versed in anatomy, which I taught for a long time," he recalls telling Armailhac in *A Doctor at Calvary: The Passion of Our Lord Jesus Christ as Described by a Surgeon*. "I lived for thirteen years in close contact with corpses," reads the next line. One assumes that the teaching stint and the years spent living in close contact with corpses were one and the same, but who knows. Perhaps he kept dead family members in the cellar.

Little is known about our Dr. Barbet, except that he became
very devoted, possibly a little too devoted, to proving the
authenticity of the Shroud. One day soon, he would find himself up in his lab, pounding nails into the hands and feet of an
elfin, Einstein-haired cadaver—one of the many unclaimed
dead brought as a matter of course to Parisian anatomy labs—
and crucifying the dead man on a cross of his own making.

Barbet had become fixated on a pair of elongated "blood-
stains"* issuing from the "imprint" of the back of the right
hand on the shroud. The two stains come from the same
source but proceed along different paths, at different angles.
The first, he writes, "mounts obliquely upwards and inwards
(anatomically its position is like that of a soldier when challenging), reaching the ulnar edge of the forearm. Another
flow, but one more slender and meandering, has gone upwards
as far as the elbow." In the soldier remark, we have an early
glimmer of what, in the due course of time, became clear:
Barbet was something of a wack. I mean, I don't wish to be
unkind, but who uses battle imagery to describe the angle of
a blood flow?

Barbet decided that the two flows were created by Jesus'
alternately pushing himself up and then sagging down to hang
by his hands; thus the trickle of blood from the nail wound
would follow two different paths, depending on which position he was in. The reason Jesus was doing this, Barbet theo-
rized, was that when people hang from their arms, it becomes
difficult to exhale; Jesus was trying to keep from suffocating.
Then, after a while, his legs would fatigue and he'd sag back

---

* Is it really blood on the Shroud of Turin? According to forensic tests done by
the late Alan Adler, a chemist and a Shroudie, it most certainly is. According to
Joe Nickell, author of *Inquest on the Shroud of Turin*, it most certainly isn't. In an
article on the Web site of the famed debunking group Committee for the Scientific Investigation of Claims of the Paranormal, Nickell says forensic tests of the
"blood" have shown it to be a mixture of red ocher and vermilion tempera paint.

down again. Barbet cited as support for his idea a torture technique used during World War I, wherein the victim is hung by his hands, which are bound together over his head. "Hanging by the hands causes a variety of cramps and contractions," wrote Barbet. "Eventually these reach the inspiratory muscles and prevent expiration; the condemned men, being unable to empty their lungs, die of asphyxia."

Barbet used the angles of the purported blood flows on the shroud to calculate what Jesus' two positions on the cross must have been: In the sagging posture, he calculated that the outstretched arms formed a 65-degree angle with the stipes (the upright beam) of the cross. In the pushed-up position, the arms formed a 70-degree angle with the stipes. Barbet then tried to verify this, using one of the many unclaimed corpses that were delivered to the anatomy department from the city's hospitals and poorhouses.

Once Barbet got the body back to his lab, he proceeded to nail it to a homemade cross. He then raised the cross upright and measured the angle of the arms when the slumping body came to a stop. Lo and behold, it was 65 degrees. (As the cadaver could of course not be persuaded to push itself back up, the second angle remained unverified.) The French edition of Barbet's book includes a photograph of the dead man on the cross. The cadaver is shown from the waist up, so I cannot say whether Barbet dressed him Jesus-style in swaddling undergarments, but I can say that he bears an uncanny resemblance to the monologuist Spalding Gray.

Barbet's idea presented an anatomical conundrum. For if there were periods when Jesus' legs gave out and he was forced to hang the entire weight of his body off his nailed palms, wouldn't the nails rip through the flesh? Barbet wondered whether, in fact, Jesus had been nailed through the stronger, bonier wrists, and not the flesh of the palms. He decided to do an experiment, detailed in *A Doctor at Calvary*. This time, rather than wrestle another whole cadaver onto his cross, he crucified a lone arm. Barely had the owner of the arm left the room when Barbet had his hammer out:

> Having just amputated an arm two-thirds of the way up from a vigorous man, I drove a square nail of about 1/3 of an inch (the nail of the Passion) into the middle of the palm. . . . I gently suspended a weight of 100 pounds from the elbow (half the weight of the body of a man about 6 foot tall). After ten minutes, the wound had lengthened; . . . I then gave the whole a moderate shake, and I saw the nail suddenly forcing its way through the space between the two metacarpal heads and making a large tear in the skin. . . . A second slight shake tore away what skin remained.

In the weeks that followed, Barbet went through twelve more arms in a quest to find a suitable point in the human wrist through which to hammer a 1/3-inch nail. This was not a good time for vigorous men with minor hand injuries to visit the offices of Dr. Pierre Barbet.

Eventually, Barbet's busy hammer made its way to what he believed was the true site of the nail's passage: Destot's space, a pea-sized gap between the two rows of the bones of the wrist. "In each case," he wrote, "the point took up its own direction and seemed to be slipping along the walls of a funnel and then to find its way spontaneously into the space which was awaiting it." It was as though divine intervention applied to nail trajectories as well. "And this spot," Barbet continued triumphantly, "is precisely where the shroud shows us the mark of the nail, a spot of which no forger would have had any idea. . . ."

And then along came Frederick Zugibe.

Zugibe is a gruff, overworked medical examiner for Rockland County, New York, who spends his spare time researching the Crucifixion and "Barbet-bashing" at what he calls "Shroudie conferences" around the world. He'll always make time to talk to you if you call, but it becomes quickly clear in the course of the conversation that spare time is something Zugibe has very little of. He'll be halfway through an explanation of the formula used to determine the pull of the body on each of Christ's hands when his voice will wander away from the telephone for a minute, and then he'll come back and say, "Excuse me. A nine-year-old body. Father beat her to death. Where were we?"

Zugibe is not on a mission to prove the authenticity of the Shroud of Turin—as, I suspect, Barbet was. He became interested in the science of crucifixion fifty years ago, as a biology student, when someone gave him a paper to read about the medical aspects of the Crucifixion. The physiological information in the paper struck him as inaccurate. "So I researched it out, wrote a term paper, got interested." The Shroud of Turin interested him only in that it might, were it for real, provide a great deal of information about the physiology of crucifixion. "Then I came across Barbet. I thought, Gee, this is exciting. Must be a real smart guy—double blood flow and all that." Zugibe began doing research of his own. One by one, Barbet's theories fell apart.

Like Barbet, Zugibe constructed a cross, which has stood—with the exception of several days during 2001 when it was out for repairs (warped stipes)—in his garage in suburban New York for some forty years. Rather than crucifying corpses, Zugibe uses live volunteers, hundreds in all. For his first study, he recruited just shy of one hundred volunteers from a local religious group, the Third Order of St. Francis. How much do you have to pay a research subject to be crucified? Nothing. "They would have paid me," says Zugibe. "Everyone wanted to go up and see what it felt like." Granted, Zugibe was using leather straps, not nails. (Over the years, Zugibe has occasionally received calls from volunteers seeking the real deal. "Would you believe? A girl called me and wanted me to actually nail her. She's with this group where they put plates in their face, they surgically alter their heads, they bifurcate their tongues and put those things through their penis.")

The first thing Zugibe noticed when he began putting people up on his cross was that none of them were having trouble breathing, even when they stayed up there for forty-five minutes. (He'd been skeptical about Barbet's suffocation theory and dismissive of the reference to torture victims because those men's hands were directly over their heads, not out to their sides.) Nor did Zugibe's subjects spontaneously try to lift themselves up. In fact, when asked to do so, in a different experiment, they were unable to. "It is totally impossible to lift yourself up from that position, with the feet flush to the cross," Zugibe asserts. Furthermore, he points out, the double blood flows were on the back of the hand, which was pressed against the cross. If Jesus had been pushing himself up and down, the blood oozing from the wound would have been smeared, not neatly split into two flows.

What, then, could have caused the famed double flow marks on the Shroud? Zugibe imagines its having happened after Jesus was taken down from the cross and washed. The washing disturbed the clotting and a small quantity of blood trickled out and split into two rivulets as it encountered the ulnar styloid protuberance, the bump that protrudes from the pinkie side of the wrist. Zugibe recalled having seen a flow of blood just like this on a gunshot victim in his lab. He tested his theory by washing the dried blood from the wound of a recently arrived corpse in his lab to see if a small quantity of blood might leak out. "Within a few minutes," he writes in an article published in the Shroudie journal *Sindon*, "a small rivulet of blood appeared."

Zugibe then noticed that Barbet had made an anatomical blunder regarding Destot's space, which is not, as Barbet crowed in his book, "precisely where the shroud shows us the mark of the nail." The wound on the back of the hand on the Shroud of Turin appears on the thumb side of the wrist, and any anatomy textbook will confirm that Destot's space is on the pinkie side of the wrist, where Barbet indeed sank his nails into his cadaver wrists.

Zugibe's theory holds that the nail went in through Jesus' palm at an angle and came out the back side at the wrist. He has his own brand of cadaveric evidence: photographs taken forty-four years ago of a murder victim that showed up in his lab. "She'd been brutally stabbed over her whole body," Zugibe recalls. "I found a defense wound where she had raised her hand in an attempt to protect her face from the vicious onslaught." Though the entry wound was in the palm, the knife had apparently traveled at an angle, coming out the back of the wrist on the thumb side. The pathway of the knife apparently offered little resistance: An X-ray showed no chipped bones.

There is a photograph of Zugibe and one of his volunteers in the aforementioned *Sindon* article. Zugibe is dressed in a knee-length white lab coat and is shown adjusting one of the vital sign leads affixed to the man's chest. The cross reaches almost to the ceiling, towering over Zugibe and his bank of medical monitors. The volunteer is naked except for a pair of gym shorts and a hearty mustache. He wears the unconcerned, mildly zoned-out expression of a person waiting at a bus stop. Neither man appears to have been self-conscious about being photographed this way. I think that when you get yourself down deep into a project like this, you lose sight of how odd you must appear to the rest of the world.

No doubt Pierre Barbet saw nothing strange or wrong in

using cadavers meant for the teaching of anatomy as subjects in a simulated crucifixion to prove to doubters that the miraculous Shroud of Turin was for real. "It is indeed essential," he wrote in the introduction to *A Doctor at Calvary*, "that we, who are doctors, anatomists, and physiologists, that we who know, should proclaim abroad the terrible truth that our poor science should no longer be used merely to alleviate the pains of our brothers, but should fulfill a greater office, that of enlightening them."

To my mind there is no "greater office" than that of "alleviating the pains of our brothers"—certainly not the office of religious propaganda. Some people, as we're about to see, manage to alleviate their brothers' pains and sufferings while utterly dead. If there were ever a cadaver eligible for sainthood, it would not be our Spalding Gray upon the cross, it would be these guys: the brain-dead, beating-heart organ donors that come and go in our hospitals every day.

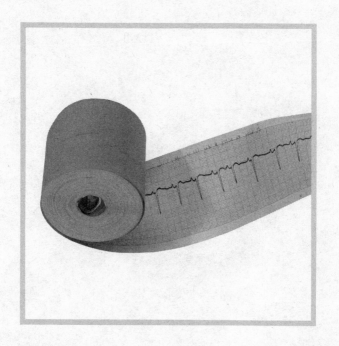

8

HOW TO KNOW IF YOU'RE DEAD

———

Beating-heart cadavers, live burial,
and the scientific search for the soul

A patient on the way to surgery travels at twice the speed of a patient on the way to the morgue. Gurneys that ferry the living through hospital corridors move forward in an aura of purpose and push, flanked by caregivers with long strides and set faces, steadying IVs, pumping ambu bags, barreling into double doors. A gurney with a cadaver commands no urgency. It is wheeled by a single person, calmly and with little notice, like a shopping cart.

For this reason, I thought I would be able to tell when the dead woman was wheeled past. I have been standing around at the nurses' station on one of the surgery floors of the University of California at San Francisco Medical Center, watching gurneys go by and waiting for Von Peterson, public affairs manager of the California Transplant Donor Network, and a cadaver I will call H. "There's your patient," says the charge nurse. A commotion of turquoise legs passes with unexpected forward-leaning urgency.

H is unique in that she is both a dead person *and* a patient on the way to surgery. She is what's known as a "beating-heart cadaver," medically functional everywhere but her brain. Up until artificial respiration was developed, there was no such entity; without a functioning brain, a body will not breathe on its own. But hook it up to a respirator and its heart will beat, and the rest of its organs will, for a matter of days, continue to thrive.

H doesn't look or smell or feel dead. If you leaned in close over the gurney, you could see her pulse beating in the arteries of her neck. If you touched her arm, you would find it warm and resilient, like your own. This is perhaps why the nurses and doctors refer to H as a patient, and why she makes her entrance to the OR at the customary presurgery clip.

Since brain death is the legal definition of death in this country, H the person is certifiably dead. But H the organs and tissues is very much alive. These two seemingly contradictory facts afford her an opportunity most corpses do not have: that of extending the lives of two or three dying strangers. Over the next four hours, H will surrender her liver, kidneys, and heart. One at a time, surgeons will come and go, taking an organ and returning in haste to their stricken patients. Until recently, the process was known among transplant professionals as an "organ harvest," which had a joyous, celebratory ring to it, perhaps a little too joyous, as it has been of late replaced by the more businesslike "organ recovery."

In H's case, one surgeon will be traveling from Utah to recover her heart, and another, the one recovering both the liver and the kidneys, will be taking them two floors down. UCSF is a major transplant center, and organs removed here often remain in house. More typically, a transplant patient's surgeon will travel from UCSF to a small town somewhere to retrieve the organ—often from an accident victim, someone young with strong, healthy organs, whose brain took an unexpected hit. The doctor does this because typically there is no doctor in that small town with experience in organ recovery. Contrary to rumors about surgically trained thugs cutting people open in hotel rooms and stealing their kidneys, organ recovery is tricky work. If you want to be sure it's done right, you get on a plane and go do it yourself.

Today's abdominal recovery surgeon is named Andy Posselt. He is holding an electric cauterizing wand, which looks like a cheap bank pen on a cord but functions like a scalpel. The wand both cuts and burns, so that as the incision is made, any vessels that are severed are simultaneously melted shut. The result is that there is a good deal less bleeding and a good deal more smoke and smell. It's not a bad smell, but simply a seared-meat sort of smell. I want to ask Dr. Posselt whether he likes it, but I can't bring myself to, so instead I ask whether he thinks it's bad that I like the smell, which I don't really, or maybe just a little. He replies that it is neither bad nor good, just morbid.

I have never before seen major surgery, only its scars. From the length of them, I had imagined surgeons doing their business, taking things out and putting them in, through an opening maybe eight or nine inches long, like a woman poking around for her glasses at the bottom of her purse. Dr. Posselt begins just above H's pubic hair and proceeds a good two feet north, to the base of her neck. He's unzipping her like a parka. Her sternum is sawed lengthwise so that her rib cage can be parted, and a large retractor is installed to pull the two sides of the incision apart so that it is now as wide as it is long. To see her this way, held open like a Gladstone bag, forces a view of the human torso for what it basically is: a large, sturdy container for guts.

On the inside, H looks very much alive. You can see the pulse of her heartbeat in her liver and all the way down her aorta. She bleeds where she is cut and her organs are plump and slippery-looking. The electronic beat of the heart monitor reinforces the impression that this is a living, breathing, thriving person. It is strange, almost impossible, really, to think of her as a corpse. When I tried to explain beating-heart cadavers to my stepdaughter Phoebe yesterday, it didn't make sense to her. But if their heart is beating, aren't they still a person? she wanted to know. In the end she decided they were "a kind of person you could play tricks on but they wouldn't know." Which, I think, is a pretty good way of summing up most donated cadavers. The things that happen to the dead in labs and ORs are like gossip passed behind one's back. They are not felt or known and so they cause no pain.

The contradictions and counterintuitions of the beating-heart cadaver can exact an emotional toll on the intensive care unit (ICU) staff, who must, in the days preceding the harvest, not only think of patients like H as living beings, but treat and care for them that way as well. The cadaver must be monitored around the clock and "life-saving" interventions undertaken on its behalf. Since the brain can no longer regulate blood pressure or the levels of hormones and their release into the bloodstream, these things must be done by ICU staff, in order to keep the organs from degrading. Observed a group of Case Western Reserve University School of Medicine physicians in a *New England Journal of Medicine* article entitled "Psychosocial and Ethical Implications of Organ Retrieval": "Intensive care unit personnel may feel confused about having to perform cardiopulmonary resuscitation on a patient who has been declared dead, whereas a 'do not resuscitate' order has been written for a living patient in the next bed."

The confusion people feel over beating-heart cadavers reflects centuries of confusion over how, exactly, to define death, to pinpoint the precise moment when the spirit—the soul, the chi, whatever you wish to call it—has ceased to exist and all that remains is a corpse. Before brain activity could be measured, the stopping of the heart had long been considered the defining moment. In point of fact, the brain survives for six to ten minutes after the heart has stopped pumping blood to it, but this is splitting hairs, and the definition works quite well for the most part. The problem, for centuries, was that doctors couldn't tell for sure whether the heart had ceased to beat or whether they were merely having trouble hearing it. The stethoscope wasn't invented until the mid-1800s, and the early models amounted to little more than a sort of medical ear trumpet. In cases where the heartbeat and pulse are especially faint—drownings, stroke, certain types of narcotic poisoning—even the most scrupulous physician had difficulty telling, and patients ran the risk of being dispatched to the undertaker before they'd actually expired.

To allay patients' considerable fears of live burial, as well as their own insecurities, eighteenth- and nineteenth-century physicians devised a diverting roster of methods for verifying death. Welsh physician and medical historian Jan Bondeson collected dozens of them for his witty and admirably researched book *Buried Alive*. The techniques seemed to fall into two categories: those that purported to rouse the unconscious patient with unspeakable pain, and those that threw in a measure of humiliation. The soles of the feet were sliced with razors, and needles jammed beneath toenails. Ears were assaulted with bugle fanfares and "hideous Shrieks and excessive Noises." One French clergyman recommended thrusting a red-hot poker up what Bondeson genteelly refers to as "the rear passage." A French physician invented a set of nipple pincers specifically for the purpose of reanimation. Another invented a bagpipelike contraption for administering tobacco enemas, which he demonstrated enthusiastically on cadavers in the morgues of Paris. The seventeenth-century anatomist Jacob Winslow entreated his colleagues to pour boiling Spanish wax on patients' foreheads and warm urine into their mouths. One Swedish tract on the matter suggested that a crawling insect be put into the corpse's ear. For simplicity and originality, though, nothing quite matches the thrusting of "a sharp pencil" up the presumed cadaver's nose.

In some cases, it is unclear who was the more humiliated, patient or doctor. French physician Jean Baptiste Vincent Laborde wrote at great length of his technique of rhythmic tongue-pulling, which was to be carried out for no less than three hours following the suspected death. (He later invented a hand-cranked tongue-pulling machine, which made the task less unpleasant though only marginally less tedious.) Another French physician instructed doctors to stick one of the patient's fingers in their ear, to listen for the buzzing sound produced by involuntary muscle movement.

Not all that surprisingly, none of these techniques gained wide acceptance, and most doctors felt that putrefaction was the only reliable way to verify that someone was dead. This meant that corpses had to sit around the house or the doctor's office for two or three days until the telltale signs and smells could be detected, a prospect perhaps even less appealing than giving them enemas. And so it was that special buildings, called waiting mortuaries, were built for the purpose of ware-housing the moldering dead. These were huge, ornate halls, common in Germany in the 1800s. Some had separate halls for male and female cadavers, as though, even in death, men couldn't be trusted to comport themselves respectably in the presence of a lady. Others were segregated by class, with the well-to-do deceased paying extra to rot in luxury surround-ings. Attendants were employed to keep watch for signs of life, which they did via a system of strings linking the fingers of corpses to a bell* or, in one case, the bellows of a large organ,

---

* I read on a Web site somewhere that this was the origin of the saying "Saved by the bell." In fact, by one reckoning, not a single corpse of the million-plus sent to waiting mortuaries over a twenty-year period awakened. If the bell alerted the attendant, which it often did, it was due to the corpse's shifting and collapsing as it decomposed. This was the origin of the saying "Driven to seek new employ-ment by the bell," which you don't hear much anymore and probably never did, because I made it up.

so that any motion on the part of the deceased would alert the attendant, who was posted, owing to the considerable stench, in a separate room. As years passed and not a single resident was saved, the establishments began to close, and by 1940, the waiting mortuary had gone the way of the nipple pincer and the tongue puller.

If only the soul could be seen as it left the body, or somehow measured. That way, determining when death had occurred would be a simple matter of scientific observation. This almost became a reality, at the hands of a Dr. Duncan Macdougall, of Haverhill, Massachusetts. In 1907, Macdougall began a series of experiments seeking to determine whether the soul could be weighed. Six dying patients, one after another, were installed on a special bed in Macdougall's office that sat upon a platform beam scale sensitive to two-tenths of an ounce. By watching for changes in the weight of a human being before, and in the act of, dying, he sought to prove that the soul had substance. Macdougall's report of the experiment was published in the April 1907 issue of *American Medicine,* considerably livening up the usual assortment of angina and urethritis papers. Below is Macdougall describing the first subject's death. He was nothing if not thorough.

At the end of three hours and forty minutes he expired and suddenly coincident with death the beam end dropped with an audible stroke hitting against the lower limiting bar and remaining there with no rebound. The loss was ascertained to be three-fourths of an ounce.

This loss of weight could not be due to evaporation of respiratory moisture and sweat, because that had already been determined to go on, in his case, at the rate of one-sixtieth of an ounce per minute, whereas this loss was sudden and large. . . .

The bowels did not move; and if they had moved the weight

would still have remained upon the bed except for a slow loss by the evaporation of moisture, depending, of course, upon the fluidity of the feces. The bladder evacuated one or two drams of urine. This remained upon the bed and could only have influenced the weight by slow gradual evaporation and therefore in no way could account for the sudden loss.

There remained but one more channel of loss to explore, the expiration of all but the residual air in the lungs. Getting upon the bed myself, my colleague put the beam at actual balance. Inspiration and expiration of air as forcibly as possible by me had no effect upon the beams. . . .

After watching another five patients shed similar weight as they died, Macdougall moved on to dogs. Fifteen dogs breathed their last without registering a significant drop in weight, which Macdougall took as corroborating evidence, for he assumed, in keeping with his religious doctrine, that animals have no souls. While Macdougall's human subjects were patients of his, there is no explanation of how he came to be in the possession of fifteen dying dogs in so short a span of time. Barring a local outbreak of distemper, one is forced to conjecture that the good doctor calmly poisoned fifteen healthy canines for his little exercise in biological theology.

Macdougall's paper sparked an acrid debate in the *American Medicine* letters column. Fellow Massachusetts doctor Augustus P. Clarke took Macdougall to task for having failed to take into account the sudden rise in body temperature at death when the blood stops being air-cooled via its circulation through the lungs. Clarke posited that the sweating and moisture evaporation caused by this rise in body temperature would account both for the drop in the men's weight and the dogs' failure to register one. (Dogs cool themselves by panting, not sweating.) Macdougall rebutted that without circulation, no blood can be brought to the surface of the skin and thus no surface cooling occurs. The debate went on from the May issue all the way through December, whereupon I lost the thread, my eye having strayed across the page to "A Few Points in the Ancient History of Medicine and Surgery," by Harry H. Grigg, M.D. It is with thanks to Harry H. Grigg that I can now hold forth at cocktail parties on the history of hemorrhoids, gonorrhea, circumcision, and the speculum.*

With improvements in stethoscopes and gains in medical knowledge, physicians began to trust themselves to be able to tell when a heart had stopped, and medical science came to agree that this was the best way to determine whether a patient had checked out for good or was merely down the hall getting ice. Placing the heart center stage in our definition of death served to give it, by proxy, a starring role in our definition of life and the soul, or spirit or self. It has long had this anyway, as evidenced by a hundred thousand love songs and sonnets and I ♥ bumper stickers. The concept of the beating-heart cadaver, grounded in a belief that the self resides in the brain and the brain alone, delivered a philosophical curveball. The notion of the heart as fuel pump took some getting used to.

The seat-of-the-soul debate has been ongoing some four thousand years. It started out not as a heart-versus-brain debate,

---

* Since the odds of our meeting at a cocktail party are slim and the odds of my managing to swing the conversation around to speculums slimmer still, let me take this opportunity to share: The earliest speculum dates from Hippocrates' day and was a rectal model. It was to be another five hundred years before the vaginal speculum made its debut. Dr. Grigg theorizes that this was because, in the Arabian model of medicine followed at the time, women could be examined only by women, and there were very few women doctors to do the examining. This implies that most women in Hippocrates' day never went to the gyno. Given that the Hippocratic gynecological cabinet included cow-dung pessaries and fumigation materials "of heavy and foul smell"—not to mention rectal speculums—they were probably better off.

but as heart-versus-liver. The ancient Egyptians were the original heart guys. They believed that the ka resided in the heart.
Ka was the essence of the person: spirit, intelligence, feelings
and passions, humor, grudges, annoying television theme
songs, all the things that make a person a person and not a
nematode. The heart was the only organ left inside a mummified corpse, for a man needed his ka in the afterlife. The brain
he clearly did not need: cadaver brains were scrambled and
pulled out in globs, through the nostrils, by way of a hooked
bronze needle. Then they were thrown away. (The liver,
stomach, intestines, and lungs were taken out of the body, but
kept: They were stored in earthen jars inside the tomb, on the
assumption, I guess, that it is better to overpack than to leave
something behind, particularly when packing for the afterlife.)

The Babylonians were the original liver guys, believing
the organ to be the source of human emotion and spirit. The
Mes-opotamians played both sides of the argument, assigning emotion to the liver and intellect to the heart. These
guys clearly marched to the beat of a freethinking drummer,
for they assigned a further portion of the soul (cunning) to
the stomach. Similar freethinkers throughout history have
included Descartes, who wrote that the soul could be found in
the walnut-sized pineal gland, and the Alexandrian anatomist
Strato, who decided it lived "behind the eyebrows."

With the rise of classical Greece, the soul debate evolved
into the more familiar heart-versus-brain, the liver having
been demoted to an accessory role.* Though Pythagorus and
Aristotle viewed the heart as the seat of the soul—the source
of "vital force" necessary to live and grow—they believed

---

* We are fortunate that this is so, for we would otherwise have been faced with
Céline Dion singing "My Liver Belongs to You" and movie houses playing *The
Liver Is a Lonely Hunter*. Every Spanish love song that contains the word *corazón,*
which is all of them, would contain the somewhat less lilting *higado,* and bumper
stickers would proclaim, "I [liver symbol] my Pekingese."

there to be a secondary, "rational" soul, or mind, located in the brain. Plato agreed that both the heart and the brain were soul terrain, but assigned primacy to the brain. Hippocrates, for his part, seemed confused (or perhaps it's me). He noted the effects of a crushed brain upon speech and intelligence, yet referred to it as a mucus-secreting gland, and wrote elsewhere that intelligence and "heat," which he said controlled the soul, were located in the heart.

The early anatomists weren't able to shed much light on the issue, as the soul wasn't something you could see or set your scalpel to. Lacking any scientific means of pinning down the soul, the first anatomists settled on generative primacy: What shows up first in the embryo must be most important and therefore most likely to hold the soul. The trouble with this particular avenue of learning, known as ensoulment, was that early first-trimester human embryos were difficult to come by. Classical scholars of ensoulment, Aristotle among them, attempted to get around the problem by examining the larger, more easily obtained poultry embryo. To quote Vivian Nutton, author of "The Anatomy of the Soul in Early Renaissance Medicine" in *The Human Embryo*, "Analogies drawn from the inspection of hen's eggs foundered on the objection that man was not a chicken."

According to Nutton, the man who came closest to actually examining a human embryo was an anatomist named Realdo Colombo, who, at the behest of the Renaissance philosopher Girolamo Pontano,* dissected a one-month-old fetus. Colombo returned from his lab—which in all likelihood was not equipped with a microscope, as the device had barely been invented—bearing the fascinating if flat-out wrong news that the liver formed before the heart.

Living amid our culture's heart-centric rhetoric, the valentines and the pop song lyrics, it is hard to imagine assigning

---

* I'd never heard of him, either.

spiritual or emotional sovereignty to the liver. Part of the reason for its exalted status among the early anatomists was that they erroneously believed it to be the origin of all the body's blood vessels. (William Harvey's discovery of the circulatory system dealt the liver-as-seat-of-the-soul theory a final fatal blow; Harvey, you will not be surprised to hear, believed that the soul was carried in the blood.) I think it was something else too. The human liver is a boss-looking organ. It's glossy, aerodynamic, Olympian. It looks like sculpture, not guts. I've been marveling at H's liver, currently being prepped for its upcoming journey. The organs around it are amorphous and unappealing. Stomachs are flappy, indistinct; intestines, chaotic and soupy. Kidneys skulk under bundles of fat. But the liver gleams. It looks engineered and carefully wrought. Its flanks have a subtle curve, like the horizon seen from space. If I were an ancient Babylonian, I guess I might think God splashed down here too.

Dr. Posselt is isolating the vessels and connectors on the liver and kidneys, prepping them for the organs' removal. The heart will go first—hearts remain viable only four to six hours; kidneys, by contrast, can be held in cold storage eighteen or even twenty-four hours—but the heart recovery surgeon hasn't arrived. He's flying in from Utah.

Minutes later a nurse puts her head through the OR doors. "Utah's in the building." People who work in ORs talk to each other in the truncated, slang-heavy manner of pilots and flight control types. The schedule on the OR wall lists today's procedure—the removal of four vital organs in preparation for death-defying transplantation into three desperate human beings—as "Recovery abdm (liv/kid x2) ♥." A few minutes ago, someone made reference to "the panky," meaning "the pancreas."

"Utah's changing."

Utah is a gentle-looking man of perhaps fifty, with graying

hair and a thin, tanned face. He has finished changing and a nurse is snapping on his gloves. He looks calm, competent, even a little bored. (This just slays me. The man is about to cut a beating heart out of a human chest.) The heart has been hidden until now behind the pericardium, a thick protective sac which Dr. Posselt now cuts away.

There is her heart. I've never seen one beating. I had no idea they moved so much. You put your hand on your heart and you picture something pulsing slightly but basically still, like a hand on a desktop tapping Morse code. This thing is going wild in there. It's a mixing-machine part, a stoat squirming in its burrow, an alien life form that's just won a Pontiac on *The Price Is Right*. If you were looking for the home of the human body's animating spirit, I could imagine believing it to be here, for the simple reason that it is the human body's most animated organ.

Utah places clamps on the arteries of H's heart, stanching the flow of blood in preparation for the cuts. You can tell by the vital signs monitor that something monumental is happening to her body. The ECG has quit drawing barbed wire and begun to look like a toddler's Etch-a-Sketch scrawls. A quick geyser of blood splashes Utah's glasses, then subsides. If H weren't dead, she'd be dying now.

This is the moment, reported the Case Western Reserve group who interviewed transplant professionals, when OR staff have been known to report sensing a "presence" or "spirit" in the room. I try to raise the mental aerial and keep myself open to the vibes. Of course I have no idea how to do this. When I was six, I tried as hard as I could to will my brother's GI Joe to walk across the room to him. This is how these extrasensory deals go with me: Nothing comes of it, and then I feel stupid for trying.

Here is the deeply unnerving thing: The heart, cut from the chest, keeps beating on its own. Did Poe know this when he wrote "The Tell-Tale Heart"? So animated are these freestanding hearts that surgeons have been known to drop them. "We wash them off and they do just fine," replied New York heart transplant surgeon Mehmet Oz when I asked him about it. I imagined the heart slipping across the linoleum, the looks exchanged, the rush to retrieve it and clean it off, like a bratwurst that's rolled off the plate in a restaurant kitchen. I ask about these things, I think, because of a need to make human what otherwise verges on the godlike: taking live organs from bodies and making them live in another body. I also asked whether the surgeons ever set aside the old, damaged hearts of transplant recipients for them to keep. Surprisingly (to me, anyway), only a few express an interest in seeing or keeping their hearts.

Oz told me that a human heart removed from its blood supply can continue beating for as long as a minute or two, until the cells begin to starve from lack of oxygen. It was phenomena like this that threw eighteenth-century medical philosophers into a tizzy: If the soul was in the brain and not the heart, as many believed at that time, how could the heart keep beating outside the body, cut off from the soul?

Robert Whytt was particularly obsessed with the matter. Beginning in 1761, Whytt was the personal physician to His Majesty the King of England, whenever His Majesty traveled north to Scotland, which wasn't all that often.* When he wasn't busy with His Majesty's bladder stones and gout, he

---

* No matter, for Whytt could have kept his appointment book full with no other patient besides himself. According to R. K. French's biography of Whytt in the Wellcome Institute of the History of Medicine series, edited by F. N. L. Poynter, M.D., the physician suffered from gout, spastic bowels, "frequent flatulence," a "disordered stomach," "wind in the stomach," nightmares, giddiness, faintness, depression, diabetes, purple discolorations of the thighs and lower legs, coughing fits "producing a thick phlegm," and, according to two of Whytt's colleagues, hypochondria. When he died, at the age of fifty-two, he was found to have "some five pounds of fluid, mixed with a substance of gelatinous consistency and bluish color," in his chest, a "red spot the size of a shilling on the mucous membrane of

could be found in his lab, cutting the hearts out of live frogs and chickens and, in one memorable instance that you hope for Whytt's sake His Majesty never got wind of, dribbling saliva onto the heart of a decapitated pigeon in an attempt to start it up again. Whytt was one of a handful of inquiring medical minds who attempted to use scientific experimentation to pin down the location and properties of the soul. You could see from his chapter on the topic in his 1751 *Works* that he wasn't inclined to come down on either side of the heart-versus-brain debate. The heart couldn't be the seat of the soul, for when Whytt cut the heart out of an eel, the remainder of the creature was able for some time to move about "with great force."

The brain also seemed an unlikely home port for the animating spirit, for animals had been observed to get on quite well for a surprising length of time without the benefit of a brain. Whytt wrote of the experiment of a man named Redi, who found that "a land tortoise, whose brain he extracted by a hole made in its skull, in the beginning of November, lived on to the middle of May following."* Whytt himself claimed to have been able, "by the influence of warmth," to keep the heart of a chick beating in its chest for two hours after its head was "clipped off with a pair of scissors." And then there was the experiment of a Dr. Kaau. Wrote Whytt: "A young cock whose head Dr. Kaau suddenly cut off . . . as he was running with great eagerness to his food, went on in a straight line 23 Rhinland feet, and would have gone farther had he not met with an obstacle which stoppt him." These were trying times for poultry.

Whytt began to suspect that the soul did not have a set rest-

---

the stomach," and concretions in the pancreas. (This is what happens when you put M.D.'s in charge of biographies.)

* What was going on in experiments like these? Hard to say. Perhaps the brain stem or spinal medulla had been left intact. Perhaps Dr. Redi, too, had his brain extracted from a hole in his skull the November past.

ing place in the body, but was instead diffused throughout. So that when you cut off a limb or took out an organ, a portion of the soul came along with it, and would serve to keep it animated for a time. That would explain why the eel's heart continued beating outside its body. And why, as Whytt wrote, citing a "well-known account," the "heart of a malefactor, which having been cut out of his body and thrown into the fire, leapt up several times to a considerable height."

Whytt probably hadn't heard of chi, but his concept of the ubiquitous soul has much in common with the centuries-old Eastern medical philosophy of circulating life energy. ("Chi" is also spelled "qi.") Chi is the stuff acupuncturists reroute with needles and unscrupulous healers claim to harness to cure cancer and knock people off their feet in front of TV cameras. Dozens of scientific studies purporting to document the effects of this circulating life energy have been done in Asia, many of them abstracted in the Qigong Research Database, which I browsed several years ago while researching a story on qi. All across China and Japan, qigong ("gong" means cultivation) healers are standing in labs, passing their palms over petri dishes of tumor cells, ulcer-plagued rats ("distance between rat and palm of hand is 40 cm"), and, in one particularly surreal bit of science, a foot-long section of human intestine. Few of these studies were done with controls, not because the researchers were lax, but because that's not traditionally how Eastern science is done.

The only Western-style peer-reviewed research attempting to prove the existence of life energy was done by an orthopedic surgeon and biomedical electronics expert named Robert Becker, who became interested in chi following Nixon's visit to China. Nixon, impressed with what he saw during a visit to a traditional Chinese clinic, had urged the National Institutes of Health to fund some studies. One of them was Becker's. Operating on the hypothesis that chi might be an electrical current separate from the pulses of the body's nervous system, Becker set about measuring transmission along some of the body's acupuncture meridians. Indeed, Becker reported, these lines transmitted current more efficiently.

Some years earlier, New Jersey's own Thomas Edison came up with another variation on the all-through-the-body concept of the soul. Edison believed that living beings were animated and controlled by "life units," smaller-than-microscopic entities that inhabited each and every cell and, upon death, evacuated the premises, floated around awhile, and eventually reassembled to animate a new personality—possibly another man, possibly an ocelot or a sea cucumber. Like other scientifically trained but mildly loopy* soul speculators, Edison strove to prove his theory through experimentation. In his *Diary and Sundry Observations*, Edison makes references to a set of plans for a "scientific apparatus" designed to communicate with these soullike agglomerations of life units. "Why should personalities in another existence or sphere waste their time working a little triangular piece of wood over a board with certain lettering on it?" he wrote, referring to the Ouija boards then in fashion among spirit mediums. Edison figured that the life-unit entities would put forth some sort of "etheric energy," and one need only amplify that energy to facilitate communication.

According to an April 1963 article in a journal called *Fate*, sent to me by Edison's tireless biographer Paul Israel, Edison

---

* People have trouble believing Thomas Edison to be a loopy individual. I offer as evidence the following passage on human memory, taken from his diaries: "We do not remember. A certain group of our little people do this for us. They live in that part of the brain which has become known as the 'fold of Broca.' . . . There may be twelve or fifteen shifts that change about and are on duty at different times like men in a factory. . . . Therefore it seems likely that remembering a thing is all a matter of getting in touch with the shift that was on duty when the recording was done."

died before his apparatus could be built, but rumors of a set of blueprints persisted for years. One fine day in 1941, the story goes, an inventor for General Electric named J. Gilbert Wright decided to use the closest approximation of Edison's machine—a séance and a medium—to contact the great inventor and ask him who had the plans. "You might try Ralph Fascht of 165 Pinehurst Avenue, New York, Bill Gunther of Consolidated Edison; his office is in the Empire State Building, or perhaps, best of all, Edith Ellis, 152 W. 58th St.," came the reply, confirming not only the persistence of personality after death but the persistence of the pocket address book.

Wright tracked down Edith Ellis, who sent him to a Commander Wynne, in Brooklyn, said to have a tracing of the blueprints. The mysterious Commander Wynne not only had the plans but claimed to have assembled and tried out the device. Alas, he could not make it work, and neither could Wright. You, too, can build one and take it for a spin, because the *Fate* article includes a carefully labeled ("aluminum trumpet," "wood plug," "aerial") drawing of the contraption. Wright and an associate, Harry Gardner, went on to invent their own device, an "ectoplasmic larynx," consisting of a microphone, a loudspeaker, a "sound box," and a cooperative medium with great quantities of patience. Wright used the "larynx" to contact Edison, who, apparently having nothing better to do with his afterlife than chat with the nutters, offered helpful tips on how to improve the machine.

While we're on the topic of supposedly straight-ahead but secretly loopy entities who've gotten hung up in the cellular soul area, let me tell you about a project funded and carried out by the U.S. Army. From 1981 to 1984, the U.S. Army's Intelligence and Security Command (INSCOM) was run by a Major General Albert N. Stubblebine III. At some point during his tenure, Stubblebine commissioned a senior aide to try to replicate an experiment done by Cleve Baxter, inventor of the lie detector, which purported to show that the cells of a human being, removed from that human being's being, were in some way still connected to, and able to communicate with, the mother ship. In the study, cells were taken from the inside of a volunteer's cheek, centrifuged, and put in a test tube. A readout from electrodes in the test tube was run through a sensor hooked up to the readout on a lie detector, which measures emotional excitation via heart rate, blood pressure, sweating, etc. (How you measure the vital signs on a slurry of cheek cells is beyond me, but this is the military and they know all manner of top-secret things.) So the volunteer was escorted to a room down the hall from his cheek cells and shown a disturbing videotape of unspecified violent scenes. The cells, it is said, registered a state of extreme agitation while their owner was watching the tape. The experiment was repeated at different distances over the course of two days. Even as far away as fifty miles, the cells felt the man's pain.

I wanted very badly to see the report of this experiment, so I called INSCOM. I was referred to a gentleman in the history section. First the historian said that INSCOM didn't keep records back that far. I didn't need any of the man's cheek cells to know he was lying. This is the U.S. government. They keep records of everything, in triplicate and from the dawn of time.

The historian explained that what General Stubblebine had been primarily interested in was not whether cells contain some sort of life unit or soul or cellular memory, but the phenomenon of remote viewing, wherein you can sit at your desk and call up images remote from you in time and space, like your missing cufflink or Iraqi ammunition depots or General Manuel Noriega's secret hideaway. (There was actually an Army Remote Viewing Team for a while; the CIA also contracted remote viewers.) When Stubblebine retired from the army he served as chairman of the board at a company called

Psi Tech, from which you can hire remote viewers to help you with all your remote-locating needs.

Forgive me. I have wandered far afield from my topic. But wherever it is that I am and however I feel about it, I know that all cheek cells belonging to me within fifty miles of here feel the same way.

The modern medical community is on the whole quite unequivocal about the brain being the seat of the soul, the chief commander of life and death. It is similarly unequivocal about the fact that people like H are, despite the hoochy-koochy going on behind their sternums, dead. We now know that the heart keeps beating on its own not because the soul is in there, but because it contains its own bioelectric power source, independent of the brain. As soon as H's heart is installed in someone else's chest and that person's blood begins to run through it, it will start beating anew—with no signals from the recipient's brain.

The legal community took a little longer than the physicians to come around to the concept of brain death. It was 1968 when the *Journal of the American Medical Association* published a paper by the Ad Hoc Committee of the Harvard Medical School to Examine the Definition of Brain Death advocating that irreversible coma be the new criterion for death, and clearing the ethical footpath for organ transplantation. It wasn't until 1974 that the law began to catch up. What forced the issue was a bizarre murder trial in Oakland, California.

The killer, Andrew Lyons, shot a man in the head in September 1973 and left him brain-dead. When Lyons's attorneys found out that the victim's family had donated his heart for transplantation, they tried to use this in Lyons's defense: If the heart was still beating at the time of surgery, they maintained, then how could it be that Lyons had killed him the day before?

They tried to convince the jury that, technically speaking, Andrew Lyons hadn't murdered the man, the organ recovery surgeon had. According to Stanford University heart transplant pioneer Norman Shumway, who testified in the case, the judge would have none of it. He informed the jury that the accepted criteria for death were those set forth by the Harvard committee, and that that should inform their decision. (Photographs of the victim's brains "oozing from his skull," to quote the *San Francisco Chronicle*, probably didn't help Lyons's case.) In the end, Lyons was convicted of murder. Based on the outcome of the case, California passed legislation making brain death the legal definition of death. Other states quickly followed suit.

Andrew Lyons's defense attorney wasn't the first person to cry murder when a transplant surgeon removed a heart from a brain-dead patient. In the earliest days of heart transplants, Shumway, the first U.S. surgeon to carry out the procedure, was continually harangued by the coroner in Santa Clara County, where he practiced. The coroner didn't accept the brain-death concept of death and threatened that if Shumway went ahead with his plans to remove a beating heart from a brain-dead person and use it to save another person's life, he would initiate murder charges. Though the coroner had no legal ground to stand on and Shumway went ahead anyway, the press gave it a vigorous chew. New York heart transplant surgeon Mehmet Oz recalls the Brooklyn district attorney around that time making the same threat. "He said he'd indict and arrest any heart transplant surgeon who went into his borough and harvested an organ."

The worry, explained Oz, was that someday someone who wasn't actually brain-dead was going to have his heart cut out. There exist certain rare medical conditions that can look, to the untrained or negligent eye, a lot like brain death, and the legal types didn't trust the medical types to get it right. To a very, very small degree, they had reason to worry. Take, for example, the condition known as "locked-in state." In one form of the disease, the nerves, from eyeballs to toes, suddenly and rather swiftly drop out of commission, with the result that the body is completely paralyzed, while the mind remains normal. The patient can hear what's being said but has no way of communicating that he's still in there, and that no, it's definitely not okay to give his organs away for transplant. In severe cases, even the muscles that contract to change the size of the pupils no longer function. This is bad news, for a common test of brain death is to shine a light in the patient's eyes to check for the reflexive contraction of the pupils. Typically, victims of locked-in state recover fully, provided no one has mistakenly wheeled them off to the OR to take out their heart.

Like the specter of live burial that plagued the French and German citizenry in the 1800s, the fear of live organ harvesting is almost completely without foundation. A simple EEG will prevent misdiagnosis of the locked-in state and conditions like it.

On a rational level, most people are comfortable with the concept of brain death and organ donation. But on an emotional level, they may have a harder time accepting it, particularly when they are being asked to accept it by a transplant counselor who would like them to okay the removal of a family member's beating heart. Fifty-four percent of families asked refuse consent. "They can't deal with the fear, however irrational, that the true end of their loved one will come when the heart is removed," says Oz. That they, in effect, will have killed him.

Even heart transplant surgeons sometimes have trouble accepting the notion that the heart is nothing more than a pump. When I asked Oz where he thought the soul resided, he said, "I'll confide in you that I don't think it's all in the brain. I have to believe that in many ways the core of our existence is in our heart." Does that mean he thinks the brain-dead patient isn't dead? "There's no question that the heart without a brain is of no value. But life and death is not a binary system." It's a continuum. It makes sense, for many reasons, to draw the legal line at brain death, but that doesn't mean it's really a line. "In between life and death is a state of near-death, or pseudo-life. And most people don't want what's in between."

If the heart of a brain-dead heart donor does contain something loftier than tissue and blood, some vestige of the spirit, then one could imagine that this vestige might travel along with the heart and set up housekeeping in the person who receives it. Oz once got a letter from a transplant patient who, shortly after receiving his new heart, began to experience what he could only imagine was some sort of contact with the consciousness of its previous owner. The patient, Michael "Med-O" Whitson, gave permission to quote the letter:

> I write all this with respect for the possibility that rather than some kind of contact with the consciousness of my donor's heart, these are merely hallucinations from the medications or my own projections. I know this is a very slippery slope. . . .
>
> What came to me in the first contact . . . was the horror of dying. The utter suddenness, shock, and surprise of it all. . . . The feeling of being ripped off and the dread of dying before your time. . . . This and two other incidents are by far the most terrifying experiences I have ever had. . . .
>
> What came to me on the second occasion was my donor's experience of having his heart being cut out of his chest and transplanted. There was a profound sense of violation by a mysterious, omnipotent outside force. . . .
>
> . . . The third episode was quite different than the previous two. This time the consciousness of my donor's heart

was in the present tense. . . . He was struggling to figure out where he was, even what he was. . . . It was as if none of your senses worked. . . . An extremely frightening aware- ness of total dislocation. . . . As if you are reaching with your hands to grasp something . . . but every time you reach forward your fingers end up only clutching thin air.

Of course, one man named Med-O does not a scientific inquiry make. A step in that direction is a study carried out in 1991 by a team of Viennese surgeons and psychiatrists. They interviewed forty-seven heart transplant patients about whether they had noticed any changes in their personality that they thought were due to the influence of the new heart and its former owner. Forty-four of the forty-seven said no, although the authors, in the Viennese psychoanalytic tradition, took pains to point out that many of these people responded to the question with hostility or jokes, which, in Freudian theory, would indicate some level of denial about the issue.

The experiences of the three patients who answered yes were decidedly more prosaic than were Whitson's. The first was a forty-five-year-old man who had received the heart of a seventeen-year-old boy and told the researchers, "I love to put on earphones and play loud music, something I never did before. A different car, a good stereo—those are my dreams now." The other two were less specific. One said simply that the person who had owned his heart had been a calm person and that these feelings of calm had been "passed on" to him; another felt that he was living two people's lives, replying to questions with "we" instead of "I," but offered no details about the newly acquired personality or what sort of music he enjoyed.

For juicy details, we must turn to Paul Pearsall, the author of a book called *The Heart's Code* (and another called *Super Marital Sex* and one called *Superimmunity*). Pearsall interviewed

140 heart transplant patients and presented quotes from five of them as evidence for the heart's "cellular memory" and its influence on recipients of donated hearts. There was the woman who got the heart of a gay robber who was shot in the back, and suddenly began dressing in a more feminine manner and getting "shooting pains" in her back. There was another rendition of the middle-aged man with a teenage male heart who now feels compelled to "crank up the stereo and play loud rock-and-roll music"—which I had quickly come to see as the urban myth of heart transplantation. My out-and-out favorite was the woman who got a prostitute's heart and suddenly began renting X-rated videos, demanding sex with her husband every night, and performing strip teases for him. Of course, if the woman knew that her new heart had come from a prostitute, this might have caused the changes in her behavior. Pearsall doesn't mention whether the woman knew of her donor's occupation (or, for that matter, whether he'd sent her a copy of *Super Marital Sex* before the interview).

Pearsall is not a doctor, or not, at least, one of the medical variety. He is a doctor of the variety that gets a Ph.D. and attaches it to his name on self-help book covers. I found his testimonials iffy as evidence of any sort of "cellular" memory, based as they are on crude and sometimes absurd stereotypes: that women become prostitutes because they want to have sex all day long, that gay men—gay robbers, no less—like to dress in feminine clothing. But bear in mind that I am, to quote item 13 of Pearsall's Heart Energy Amplitude Test, "cynical and distrusting of others' motives."

Mehmet Oz, the transplant surgeon I spoke with, also got curious about the phenomenon of heart transplant patients' claiming to experience memories belonging to their donors. "There was this one fellow," he told me, "who said, 'I know who gave me this heart.' He gave me a detailed description of a young black woman who died in a car accident. 'I see myself in the mirror with blood on my face and I taste French fries in my mouth. I see that I'm black and I was in this accident.' It spooked me," says Oz, "and so I went back and checked. The donor was an elderly white male." Did he have other patients who claimed to experience their donor's memories or to know something specific about their donor's life? He did. "They're all wrong."

After I spoke to Oz, I tracked down three more articles on the psychological consequences of having someone else's heart stitched into your chest. Fully half of all transplant patients, I found out, develop postoperative psychological problems of some sort. Rausch and Kneen described a man utterly terrified by the prospect of the transplant surgery, fearing that in giving up his heart he would lose his soul. Another paper presented the case of a patient who became convinced that he had been given a hen's heart. No mention was made of why he might have come to believe this or whether he had been exposed to the writings of Robert Whytt, which actually might have provided some solace, pointing out, as they do, that a chicken heart can be made to beat on for several hours in the event of decapitation—always a plus.

The worry that one will take on traits of the heart donor is quite common, particularly when patients have received, or think that they have, a heart from a donor of a different gender or sexual orientation. According to a paper by James Tabler and Robert Frierson, recipients often wonder whether the donor "was promiscuous or oversexed, homosexual or bisexual, excessively masculine or feminine or afflicted with some sort of sexual dysfunction." They spoke to a man who fantasized that his donor had had a sexual "reputation" and said he had no choice but to live up to it. Rausch and Kneen describe a forty-two-year-old firefighter who worried that his new heart, which had belonged to a woman, would make him less masculine and that his firehouse buddies would no longer accept him. (A male heart, Oz says, is in fact slightly different from a female heart. A heart surgeon can tell one from the other by looking at the ECG, because the intervals are slightly different. When you put a female heart into a man, it will continue to beat like a female heart. And vice versa.)

From reading a paper by Kraft, it would seem that when men believe their new hearts came from another man, they often believe this man to have been a stud and that some measure of this studliness has somehow been imparted to them. Nurses on transplant wards often remark that male transplant patients show a renewed interest in sex. One reported that a patient asked her to wear "something other than that shapeless scrub so he could see her breasts." A post-op who had been impotent for seven years before the operation was found holding his penis and demonstrating an erection. Another nurse spoke of a man who left the fly of his pajamas unfastened to show her his penis. Conclude Tabler and Frierson, "This irrational but common belief that the recipient will somehow develop characteristics of the donor is generally transitory but may alter sexual patterns. . . ." Let us hope that the man with the chicken heart was blessed with a patient and open-minded spouse.

The harvesting of H is winding down. The last organs to be taken, the kidneys, are being brought up and separated from the depths of her open torso. Her thorax and abdomen are filled with crushed ice, turned red from blood. "Cherry Sno-Kone," I write in my notepad. It's been almost four hours now, and H has begun to look more like a conventional cadaver, her skin dried and dulled at the edges of the incision.

The kidneys are placed in a blue plastic bowl with ice and perfusion fluid. A relief surgeon arrives for the final step of the recovery, cutting off pieces of veins and arteries to be included, like spare sweater buttons, along with the organs, in case the ones attached to them are too short to work with. A half hour later, the relief surgeon steps aside and the resident comes over to sew H up.

As he talks to Dr. Posselt about the stitching, the resident strokes the bank of fat along H's incision with his gloved hand, then pats it twice, as though comforting her. When he turns back to his work, I ask him if it feels different to be working on a dead patient.

"Oh, yes," he answers. "I mean, I would never use this kind of stitch." He has begun stitching more widely spaced, comparatively crude loops, rather than the tight, hidden stitches used on the living.

I rephrase the question: Does it feel odd to perform surgery on someone who isn't alive?

His answer is surprising. "The patient *was* alive." I suppose surgeons are used to thinking about patients—particularly ones they've never met—as no more than what they see of them: open plots of organs. And as far as that goes, I guess you could say H *was* alive. Because of the cloths covering all but her opened torso, the young man never saw her face, didn't know if she was male or female.

While the resident sews, a nurse picks stray danglies of skin and fat off the operating table with a pair of tongs and drops them inside the body cavity, as though H were a handy wastebasket. The nurse explains that this is done intentionally: "Anything not donated stays with her." The jigsaw puzzle put back in its box.

The incision is complete, and a nurse washes H off and covers her with a blanket for the trip to the morgue. Out of habit or respect, he chooses a fresh one. The transplant coordinator, Von, and the nurse lift H onto a gurney. Von wheels H into an elevator and down a hallway to the morgue. The workers are behind a set of swinging doors, in a back room.

"Can we leave this here?" Von shouts. H has become a "this."
We are instructed to wheel the gurney into the cooler, where
it joins five others. H appears no different from the corpses
already here.*

But H *is* different. She has made three sick people well. She
has brought them extra time on earth. To be able, as a dead
person, to make a gift of this magnitude is phenomenal. Most
people don't manage this sort of thing while they're alive.
Cadavers like H are the dead's heroes.

It is astounding to me, and achingly sad, that with eighty
thousand people on the waiting list for donated hearts and livers and kidneys, with sixteen a day dying there on that list, that
more than half of the people in the position H's family was in
will say no, will choose to burn those organs or let them rot.
We abide the surgeon's scalpel to save our own lives, our loved
ones' lives, but not to save a stranger's life. H has no heart, but
heartless is the last thing you'd call her.

---

* Unless H's family is planning a naked open-casket service, no one at her funeral
will be able to tell she's had organs removed. Only with tissue harvesting, which
often includes leg and arm bones, does the body take on a slightly altered profile,
and in this case PVC piping or dowels are inserted to normalize the form and
make life easier for mortuary staff and others who need to move the otherwise
somewhat noodle-ized body.

9

JUST A HEAD

———

Decapitation, reanimation, and the human head transplant

If you really wanted to know for sure that the human soul resides in the brain, you could cut off a man's head and ask it. You would have to ask quickly, for the human brain cut off from its blood supply will slide into unconsciousness after ten or twelve seconds. You would, further, have to instruct the man to answer with blinks, for, having been divorced from his lungs, he can pull no air through his larynx and thus can no longer speak. But it could be done. And if the man seemed more or less the same individual he was before you cut off his head, perhaps a little less calm, then you would know that indeed the self is there in the brain.

In Paris, in 1795, an experiment very much like this was nearly undertaken. Four years before, the guillotine had replaced the noose as the executioner's official tool. The device was named after Dr. Joseph Ignace Guillotin, though he did not invent it. He merely lobbied for its use, on the grounds that the decapitating machine, as he preferred to call it, was an instantaneous, and thus more humane, way to kill.

And then he read this:

Do you know that it is not at all certain when a head is severed from the body by the guillotine that the feelings, personality and ego are instantaneously abolished . . . ? Don't you know that the seat of the feelings and appreciation is in the brain, that this seat of consciousness can continue to operate even when the circulation of the blood is cut off from the brain . . . ? Thus, for as long as the brain retains its vital force the victim is aware of his existence. Remember that Haller insists that a head, having been removed from the shoulders of a man, grimaced horribly when a surgeon who was present stuck a finger into the rachidian canal. . . . Furthermore, credible witnesses have assured me that they have seen the teeth grind after the head has been separated from the trunk. And I am convinced that if the air could still circulate through the organs of the voice . . . these heads would speak. . . .

. . . The guillotine is a terrible torture! We must return to hanging.

It was a letter, published in the November 9, 1795, Paris *Moniteur* (and reprinted in André Soubiran's biography of Guillotin), written by the well-respected German anatomist S. T. Sömmering. Guillotin was horrified, the Paris medical community atwitter. Jean-Joseph Sue, the librarian at the Paris School of Medicine, came out in agreement with Sömmering, declaring his belief that the heads could see hear, smell, see, and think. He tried to convince his colleagues to undertake an experiment whereby "before the butchery of the victim," a few of the unfortunate's friends would arrange a code of eyelid or jaw movements which the head could use after the execution to indicate whether it was "fully conscious of [its] agony." Sue's colleagues in the medical community dismissed his idea as ghastly and absurd, and the experiment was not carried out. Nonetheless, the notion of the living head had made its way into the public consciousness and even popular literature. Below is a conversation between a pair of fictional executioners, in Alexandre Dumas's *Mille et Un Phantomes:*

"Do you believe they're dead because they've been guillotined?"

"Undoubtedly!"

"Well, one can see that you don't look in the basket when they are all there together. You've never seen them twist their eyes and grind their teeth for a good five minutes after the execution. We are forced to change the basket every three months because they cause such damage to the bottom."

Shortly after Sömmering's and Sue's pronouncements, Georges Martin, an assistant to the official Paris executioner and witness to some 120 beheadings, was interviewed on the subject of the heads and their post-execution activities. Soubiran writes that he cast his lot (not surprisingly) on the side of instantaneous death. He claimed to have viewed all 120 heads within two seconds and always "the eyes were fixed. . . . The immobility of the lids was total. The lips were already white. . . ." Medical science was, for the moment, reassured, and the furor dissipated.

But French science was not through with heads. A physiologist named Legallois surmised in an 1812 paper that if the personality did indeed reside in the brain, it should be possible to revive *une tête séparée du tronc* by giving it an injection of oxygenated blood through its severed cerebral arteries. "If a physiologist attempted this experiment on the head of a guillotined man a few instants after death," wrote Legallois's colleague Professor Vulpian, "he would perhaps bear witness to a terrible sight." Theoretically, for as long as the blood supply lasted, the head would be able to think, hear, see, smell (grind its teeth, twist its eyes, chew up the lab table), for all the nerves above the neck would still be intact and attached to the organs and muscles of the head. The head wouldn't be able to speak, owing to the aforementioned disabling of the larynx, but this was probably, from the perspective of the experimenter, just as well. Legallois lacked either the resources or the intestinal fortitude to follow through with the actual experiment, but other researchers did not.

In 1857, the French physician Brown-Séquard cut the head off a dog ("*Je décapitai un chien . . .*") to see if he could put it back in action with arterial injections of oxygenated blood. Eight minutes after the head parted company with the neck, the injections began. Two or three minutes later, Brown-Séquard noted movements of the eyes and facial muscles that appeared to him to be voluntarily directed. Clearly something was going on in the animal's brain.

With the steady supply of guillotined heads in Paris, it was only a matter of time before someone tried this out on a human. There could be only one man for the job, a man who would more than once make a name for himself (lots of names, probably) by doing peculiar things to bodies with the aim of resuscitating them. The man for the job was Jean Baptiste Vincent Laborde, the very same Jean Baptiste Vincent Laborde who appeared earlier in these pages advocating prolonged tongue-pulling as a means of reviving the comatose, mistaken-for-dead patient. In 1884, the French authorities began supplying Laborde with the heads of guillotined prisoners so that he could examine the state of their brain and nervous system. (Reports of these experiments appeared in various French medical journals, *Revue Scientifique* being the main one.) It was hoped that Laborde would get to the bottom of what he called *la terrible legende*—that it was possible for guillotined heads to be aware, if only for a moment, of their situation (in a basket, without a body). Upon a head's arrival in his lab, he would quickly bore holes in the skull and insert needles into the brain in an attempt to trigger nervous system responses. Following Brown-Séquard's lead, he also tried resuscitating the heads with a supply of blood.

Laborde's first subject was a murderer named Campi. From Laborde's description, he was not a typical thug. He had delicate ankles and white, well-manicured hands. His skin was unblemished save for an abrasion on the left cheek, which Laborde surmised was the result of the head's drop into the guillotine basket. Laborde didn't typically spend so much time personalizing his subjects, preferring to call them simply *restes frais*. The term means, literally, "fresh remains," though in French it has a pleasant culinary lilt, like something you might order off the specials board at the neighborhood bistro.

Campi arrived in two pieces, and he arrived late. Under ideal circumstances, the distance from the scaffold to Laborde's lab on Rue Vauquelin could be covered in about seven minutes. Campi's commute took an hour and twenty minutes, owing to what Laborde called "that stupid law" forbidding scientists to take possession of the remains of executed criminals until the bodies had crossed the threshold of the city cemetery. This meant Laborde's driver had to follow the heads as they "made the sentimental journey to the turnip field" (if my French serves) and then pack them up and bring them all the way back across town to the lab. Needless to say, Campi's brain had long since ceased to function in anything close to a normal state.

Infuriated by the waste of eighty critical postmortem minutes, Laborde decided to meet his next head at the cemetery gates and set directly to work on it. He and his assistants rigged a makeshift traveling laboratory in the back of a horse-drawn van, complete with lab table, five stools, candles, and the necessary equipment. The second subject was named Gamahut, a fact unlikely to be forgotten, owing to the man's having had his name tattooed on his torso. Eerily, as though presaging his gory fate, he had also been tattooed with a portrait of himself from the neck up, which, without the lines of a frame to suggest an unseen body, gave him the appearance of a floating head.

Within minutes of its arrival in the van, Gamahut's head

was installed in a styptic-lined container and the men set to work, drilling holes in the skull and inserting needles into various regions of the brain to see if they could coax any activity out of the criminal's moribund nervous system. The ability to perform brain surgery while traveling full tilt on a cobblestone street is a testament to the steadiness of Laborde's hand and/ or the craftsmanship of nineteenth-century broughams. Had the vehicle's manufacturers known, they might have crafted a persuasive ad campaign, à la the diamond cutter in the backseat of the smooth-riding Oldsmobile.

Laborde's team ran current through the needles, and the Gamahut head could be seen to make the predictable twitches of lip and jaw. At one point—to the astonished shouts of all present—the prisoner slowly opened one eye, as if, with great and understandable trepidation, he sought to figure out where he was and what sort of strange locality hell had turned out to be. But, of course, given the amount of time that had elapsed, the movement could have been nothing beyond a primitive reflex.

The third time around, Laborde resorted to basic bribery to expedite his head deliveries. With the help of the local municipality chief, the third head, that of a man named Gagny, was delivered to his lab just shy of seven minutes after the chop. The arteries on the right side of the neck were injected with oxygenated cow's blood, and, in a break from Brown-Séquard's protocol, the arteries on the other side were connected to those of a living animal: *un chien vigoureux*. Laborde had an arresting flair for details, which the medical journals of his day seemed pleased to accommodate. He devoted a full paragraph to an artful description of a severed head resting upright on the lab table, rocking ever so slightly left and right from the pulsing pressure of the dog's blood as it pumped into the head. In another paper, he took pains to detail the postmortem contents of Gamahut's excretory organs, though the information bore no relation to the experiment at hand, noting with seeming fascination that the stomach and intestines were completely empty save for *un petit bouchon fécal* at the far end.

With the Gagny head, Laborde came closest to restoring normal brain function. Muscles on the eyelids, forehead, and jaw could be made to contract. At one point Gagny's jaw snapped shut so forcefully that a loud *claquement dentaire* was heard. However, given that twenty minutes had passed from the drop of the blade to the infusion of blood—and irreversible brain death sets in after six to ten minutes—it is certain that Gagny's brain was too far gone to be brought around to anything resembling consciousness and he remained blessedly ignorant of his dismaying state of affairs. The *chien*, on the other hand, spent its final, decidedly less *vigoureux* minutes watching its blood pump into someone else's head and no doubt produced some *claquements dentaires* of its own.

Laborde soon lost interest in heads, but a team of French experimenters named Hayem and Barrier took up where he left off. The two became something of a cottage industry, trans-fusing a total of twenty-two dog heads, using blood from live horses and dogs. They built a tabletop guillotine specially fitted to the canine neck and published papers on the three phases of neurological activity following decapitation. Monsieur Guillotin would have been deeply chagrined to read the concluding statements in Hayem and Barrier's description of the initial, or "convulsive," postdecapitation phase. The physiognomy of the head, they wrote, expresses surprise or "*une grande anxiété,*" and appears to be conscious of the exterior world for three or four seconds.

Eighteen years later, a French physician by the name of Beaurieux confirmed Hayem and Barrier's observations—and Sömmering's suspicions. Using Paris's public scaffold as his lab, he carried out a series of simple observations and experiments on the head of a prisoner named Languille, the instant after the guillotine blade dropped.

Here, then, is what I was able to note immediately after the decapitation: the eyelids and lips of the guillotined man worked in irregularly rhythmic contractions for about five or six seconds . . . [and] ceased. The face relaxed, the lids half closed on the eyeballs, . . . exactly as in the dying whom we have occasion to see every day in the exercise of our profession. . . . It was then that I called in a strong, sharp voice, "Languille!" I then saw the eyelids slowly lift up, without any spasmodic contraction . . . such as happens in everyday life, with people awakened or torn from their thoughts. Next Languille's eyes very definitely fixed themselves on mine and the pupils focused themselves. I was not, then, dealing with the sort of vague dull look without any expression that can be observed any day in dying people to whom one speaks. I was dealing with undeniably living eyes which were looking at me.

After several seconds, the eyelids closed again, slowly and evenly, and the head took on the same appearance as it had had before I called out. It was at that point that I called out again, and, once more, without any spasm, slowly, the eyelids lifted and undeniably living eyes fixed themselves on mine with perhaps even more penetration than the first time. . . . I attempted the effect of a third call; there was no further movement—and the eyes took on the glazed look which they have in the dead. . . .

You know, of course, where this is leading. It is leading toward human head transplants. If a brain—a personality—and its surrounding head can be kept functional with an outside blood supply for as long as that supply lasts, then why not go the whole hog and actually transplant it onto a living, breathing body, so that it has an ongoing blood supply? Here the pages fly from the calendar and the globe spins on its stand, and we find ourselves in St. Louis, Missouri, May 1908.

Charles Guthrie was a pioneer in the field of organ transplantation. He and a colleague, Alexis Carrel, were the first to master the art of anastomosis: the stitching of one vessel to another without leaks. In those days, the task required great patience and dexterity, and very thin thread (at one point, Guthrie tried sewing with human hair). Having mastered the skill, Guthrie and Carrel went anastomosis-happy, transplanting pieces of dog thighs and entire forelimbs, keeping extra kidneys alive outside of bodies and stitching them into groins. Carrel went on to win the Nobel Prize for his contributions to medicine; Guthrie, the meeker and humbler of the two, was rudely overlooked.

On May 21, Guthrie succeeded in grafting one dog's head onto the side of another's neck, creating the world's first man-made two-headed dog. The arteries were grafted together such that the blood of the intact dog flowed through the head of the decapitated dog and then back into the intact dog's neck, where it proceeded to the brain and back into circulation. Guthrie's book *Blood Vessel Surgery and Its Applications* includes a photograph of the historic creature. Were it not for the caption, the photo would seem to be of some rare form of marsupial dog, with a large baby's head protruding from a pouch in its mother's fur. The transplanted head was sewn on at the base of the neck, upside down, so that the two dogs are chin to chin, giving an impression of intimacy, despite what must have been at the very least a strained coexistence. I imagine photographs of Guthrie and Carrel around that time having much the same quality.

As with Monsieur Gagny's head, too much time (twenty minutes) had elapsed between the beheading and the moment circulation was restored for the dog head and brain to regain much function. Guthrie recorded a series of primitive move-ments and basic reflexes, similar to what Laborde and Hayem had observed: pupil contractions, nostril twitchings, "boiling movements" of the tongue. Only one notation in Guthrie's lab notes gives the impression that the upside-down dog head might have had an awareness of what had taken place: "5:31: Secretion of tears. . . ." Both dogs were euthanized when complications set in, about seven hours after the operation.

The first dog heads to enjoy, if that word can be used, full cerebral function were those of transplantation whiz Vladi-mir Demikhov, in the Soviet Union in the 1950s. Demikhov minimized the time that the severed donor head was without oxygen by using "blood-vessel sewing machines." He trans-planted twenty puppy heads—actually, head-shoulders-lungs-and-forelimbs units with an esophagus that emptied, untidily, onto the outside of the dog—onto fully grown dogs, to see what they'd do and how long they'd last (usually from two to six days, but in one case as long as twenty-nine days).

In his book *Experimental Transplantation of Vital Organs*, Demikhov includes photographs of, and lab notes from, Experiment No. 2, on February 24, 1954: the transplantation of a one-month-old puppy's head and forelimbs to the neck of what appears to be a Siberian husky. The notes portray a lively, puppylike, if not altogether joyous existence on the part of the head:

09:00. The donor's head eagerly drank water or milk, and tugged as if trying to separate itself from the recipient's body.

22:30. When the recipient was put to bed, the transplanted head bit the finger of a member of the staff until it bled.

February 26, 18:00. The donor's head bit the recipient behind the ear, so that the latter yelped and shook its head.

Demikhov's transplant subjects were typically done in by immune reactions. Immunosuppressive drugs weren't yet available, and the immune system of the intact dog would, understandably enough, treat the dog parts grafted to its neck as a hostile invader and proceed accordingly. And so Demikhov hit a wall. Having transplanted virtually every piece and combination of pieces of a dog into or onto another dog,[*] he closed up his lab and disappeared into obscurity.

If Demikhov had known more about immunology, his career might have gone quite differently. He might have realized that the brain enjoys what is known as "immunological privilege," and can be kept alive on another body's blood supply for weeks without rejection. Because it is protected by the blood brain barrier, it isn't rejected the way other organs and tissues are. While the mucosal tissues of Guthrie's and Demikhov's transplanted dog heads began swelling and hemorrhaging within a day or two of the operation, the brains at autopsy appeared normal.

Here is where it begins to get strange.

In the mid-1960s, a neurosurgeon named Robert White began experimenting with "isolated brain preparations": a living brain taken out of one animal, hooked up to another animal's circulatory system, and kept alive. Unlike Demikhov's and Guthrie's whole head transplants, these brains, lacking faces and sensory organs, would live a life confined to memory and thought. Given that many of these dogs' and mon-

---

[*] When he tired of moving organs and heads around, Demikhov moved on to entire dog halves. His book details an operation in which two dogs were split at the diaphragm, their upper and lower halves swapped, and their arteries grafted back together. He explained that this might be less time-consuming than transplanting two or three individual organs. Given that the patient's spinal nerves, once severed, could not be reconnected and the lower half of the body would be paralyzed, the procedure failed to generate much enthusiasm.

keys' brains were implanted inside the necks and abdomens of other animals, this could only have been a blessing. While the inside of someone else's abdomen is of moderate interest in a sort of curiosity-seeking, Surgery Channel sort of way, it's not the sort of place you want to settle down in to live out the remainder of your years.

White figured out that by cooling the brain during the procedure to slow the processes by which cellular damage occurs—a technique used today in organ recovery and transplant operations—it was possible to retain most of the organ's normal functions. Which means that the personality—the psyche, the spirit, the soul—of those monkeys continued to exist, for days on end, without its body or any of its senses, inside another animal. What must that have been like? What could possibly be the purpose, the justification? Had White been thinking of one day isolating a human brain like this? What kind of person comes up with a plan like this and carries it out?

To find out, I decided to go visit White in Cleveland, where he is spending his retirement. We planned to meet at the Metro Health Care Center, downstairs from the lab where he carried out his historic operations, which has been preserved as a kind of shrine-cum-media-photo-op. I was an hour early, and spent the time driving up and down Metro Health Care Drive, looking for a place to sit and have some coffee and review White's papers. There was nothing. I ended up back at the hospital, on a patch of grass outside the parking garage. I had heard Cleveland had undergone some sort of renaissance, but apparently it underwent it in some other part of town. Let's just say it wasn't the sort of place I'd want to live out the remainder of my years, though it beats a monkey abdomen, and you can't say that about some neighborhoods.

White escorts me through the hospital corridors and stairways, past the neurosurgery department, up the stairs, to his old lab. He is seventy-six now, thinner than he was at the time of the operations, but elsewise little changed by age. His answers have the rote, patient air you expect from a man who has been asked the same questions a hundred times.

"Here we are," says White. NEUROLOGICAL RESEARCH LAB-ORATORY, says a plaque beside the door, giving away nothing. To step inside is to step back into 1968, before labs went white and stainless. The counters are of a dull black stone, stained with white rings, and the cabinets and drawers are wood. It has been a while since anyone dusted, and ivy has grown over the one window. The fluorescent lights have those old covers that look like ice tray dividers.

"This is where we shouted 'Eureka!' and danced around," recalls White. There isn't much room for dancing. It's a small, cluttered, low-ceilinged room, with a couple of stools for the scientists, and a downsized veterinary operating table for the rhesus monkeys.

And while White and his colleagues danced, what was going on inside the brain of that monkey? I ask him what he imagined it must have been like to find yourself, suddenly, reduced to your thoughts. I am, of course, not the first journalist to have asked this question. The legendary Oriana Fallaci* asked it of White's neurophysiologist Leo Massopust, in a *Look* magazine interview in November 1967. "I suspect that without his senses he can think more quickly," Dr. Massopust answered brightly. "What kind of thinking, I don't know. I guess he's primarily a memory, a repository for information stored when he had his flesh; he cannot develop further

---

* Legendary for skewering heads of state, from Kissinger to Arafat ("a man born to irritate"), Fallaci stuck it to White by making up a name for the anonymous lab monkey whose brain she had watched being isolated and for writing things like this: "While [the brain removal and hookup] happened, no one paid any attention to Libby's body, which was lying lifeless. Professor White might have fed it, too, with blood, and made it survive without a head. But Professor White didn't choose to, and so the body lay there, forgotten."

because he no longer has the nourishment of experience. Yet this, too, is a new experience."

White declines to sugar-coat. He mentions the isolation chamber studies of the 1970s, wherein subjects had no sensory input, nothing to hear, see, smell, feel, or taste. These people got as close as you can come, without White's aid, to being brains in a box. "People [in these conditions] have gone literally crazy," says White, "and it doesn't take all that long." Although insanity, too, is a new experience for most people, no one was likely to volunteer to become one of White's isolated brains. And of course, White couldn't force anyone to do it—though I imagine Oriana Fallaci came to mind. "Besides," says White, "I would question the scientific applicability. What would justify it?"

So what justified putting a rhesus monkey through it? It turns out the isolated brain experiments were simply a step on the way toward keeping entire heads alive on new bodies. By the time White appeared on the scene, early immunosuppressive drugs were available and many of the problems of tissue rejection were being resolved. If White and his team worked out the kinks with the brains and found they could be kept functioning, then they would move on to whole heads. First monkey heads, and then, they hoped, human ones.

Our conversation has moved from White's lab to a booth in a nearby Middle Eastern restaurant. My recommendation to you is that you never eat baba ganoush or, for that matter, any soft, glistening gray food item while carrying on a conversation involving monkey brains.

White thinks of the operation not as a head transplant, but as a whole-body transplant. Think of it this way: Instead of getting one or two donated organs, a dying recipient gets the entire body of a brain-dead beating-heart cadaver. Unlike Guthrie and Demikhov with their multiheaded monsters, White would remove the body donor's head and put the new one in its place. The logical recipient of this new body, as White envisions it, would be a quadriplegic. For one thing, White said, the life span of quadriplegics is typically reduced, their organs giving out more quickly than is normal. By putting them—their heads—onto new bodies, you would buy them a decade or two of life, without, in their case, much altering their quality of life. High-level quadriplegics are paralyzed from the neck down and require artificial respiration, but everything from the neck up works fine. Ditto the transplanted head. Because no neurosurgeon can yet reconnect severed spinal nerves, the person would still be a quadriplegic—but no longer one with a death sentence. "The head could hear, taste, see," says White. "It could read, and hear music. And the neck can be instrumented just like Mr. Reeve's is, to speak."

In 1971, White achieved the unthinkable. He cut the head off one monkey and connected it to the base of the neck of a second, decapitated monkey. The operation lasted eight hours and required numerous assistants, each having been given detailed instructions, including where to stand and what to say. White went up to the operating room for weeks beforehand and marked off everyone's position on the floor with chalk circles and arrows, like a football coach. The first step was to give the monkeys tracheotomies and hook them up to respirators, for their windpipes were about to severed. Next White pared the two monkey's necks down to just the spine and the main blood vessels—the two carotid arteries carrying blood to the brain and the two jugular veins bringing it back to the heart. Then he whittled down the bone on the top of the body donor's neck and capped it with a metal plate, and did the same thing on the bottom of the head. (After the vessels were reconnected, the two plates were screwed together.) Then, using long, flexible tubing, he brought the circulation of the donor body over to supply its new head and sutured the vessels. Finally, the head was cut off from the blood supply of its old body.

This is, of course, grossly simplified. I make it sound as though the whole thing could be done with a jackknife and a sewing kit. For more details, I would direct you to the July 1971 issue of *Surgery*, which contains White's paper on the procedure, complete with pen-and-ink illustrations. My favorite illustration shows a monkey body with a faint, ghostly head above its shoulders, indicating where its head had until recently been located, and a jaunty arrow arcing across the drawing toward the space above a second monkey body, where the first monkey's head is now situated. The drawing lends a tidy, businesslike neutrality to what must have been a chaotic and exceptionally gruesome operation, much the way airplane emergency exit cards give an orderly, workaday air to the interiors of crashing planes. White filmed the operation but wouldn't, despite protracted begging and wheedling, show me the film. He said it was too bloody.

That's not what would have gotten to me. What would have gotten to me was the look on the monkey's face when the anesthesia wore off and it realized what had just taken place. White described this moment in the aforementioned paper, "Cephalic Exchange Transplantation in the Monkey": "Each cephalon [head] gave evidence of the external environment. . . . The eyes tracked the movement of individuals and objects brought into their visual fields, and the cephalons remained basically pugnacious in their attitudes, as demonstrated by their biting if orally stimulated." When White placed food in their mouths, they chewed it and attempted to swallow it—a bit of a dirty trick, given that the esophagus hadn't been reconnected and was now a dead end. The monkeys lived anywhere from six hours to three days, most of them dying from rejection issues or from bleeding. (In order to prevent clotting in the anastomosed arteries, the animals were on anticoagulants, which created their own problems.)

I asked White whether any humans had ever stepped forward to volunteer their heads. He mentioned a wealthy, elderly quadriplegic in Cleveland who had made it clear that should the body transplant surgery be perfected when his time draws near, he's game to give it a whirl. "Perfected" being the key word. The trouble with human subjects is that no one wants to go first. No one wants to be a practice head.

If someone did agree to it, would White do it?

"Of course. I see no reason why it wouldn't be successful with a man." White doesn't think the United States will be the likely site of the first human head transplant, owing to the amount of bureaucracy and institutional resistance faced by inventors of radical new procedures. "You're dealing with an operation that is totally revolutionary. People can't make up their minds whether it's a total body transplant or a head transplant, a brain or even a soul transplant. There's another issue too. People will say, 'Look at all the people's lives you could save with the organs in one body, and you want to give that body to just one person. And he's *paralyzed*.'"

There are other countries, countries with less meddlesome regulating bodies, that would love to have White come over and make history swapping heads. "I could do it in Kiev tomorrow. And they're even more interested in Germany and England. And the Dominican Republic. They want me to do it. Italy would like me to do it. But where's the money?" Even in the United States, cost stands in the way: As White points out, "Who's going to fund the research when the operation is so expensive and would only benefit a small number of patients?"

Let's say someone did fund the research, and that White's procedures were streamlined and proved viable. Could there come a day when people whose bodies are succumbing to fatal diseases will simply get a new body and add decades to their lives—albeit, to quote White, as a head on a pillow? There could. Not only that, but with progress in repairing damaged spinal cords, surgeons may one day be able to reattach spinal nerves, meaning these heads could get up off their pillows and begin to move and control their new bodies. There's no reason to think it couldn't one day happen.

And few reasons to think it will. Insurance companies are unlikely to ever cover such an expensive operation, which would put this particular form of life extension out of reach of anyone but the very rich. Is it a sensible use of medical resources to keep terminally ill and extravagantly wealthy people alive? Shouldn't we, as a culture, encourage a saner, more accepting attitude toward death? White doesn't profess to have the last word on the matter. But he'd still like to do it.

Interestingly, White, a devout Catholic, is a member of the Pontifical Academy of Sciences, some seventy-eight well-known scientific minds (and their bodies) who fly to Vatican City every two years to keep the Pope up to date on scientific matters of special interest to the church: stem cell research, cloning, euthanasia, even life on other planets. In one sense, this is an odd place for White, given that Catholicism preaches that the soul occupies the whole body, not just the brain. The subject came up during one of White's meetings with the Holy Father. "I said to him, 'Well, Your Holiness, I seriously have to consider that the human spirit or soul is physically located in the brain.' The Pope looked very strained and did not answer." White stops and looks down at his coffee mug, as though perhaps regretting his candor that day.

"The Pope always looks a little strained," I point out helpfully. "I mean, with his health and all." I wonder aloud whether the Pope might be a good candidate for total body transplant. "God knows the Vatican's got the money. . . ." White throws me a look. The look says it might not be a good idea to tell White about my collection of news photographs of the Pope having trouble with his vestments. It says I'm a *petit bouchon fécal*.

White would very much like to see the church change its definition of death from "the moment the soul leaves the body" to "the moment the soul leaves the brain," especially given that Catholicism accepts both the concept of brain death and the practice of organ transplantation. But the Holy See, like White's transplanted monkey heads, has remained pugnacious in its attitude.

No matter how far the science of whole body transplantation advances, White or anyone else who chooses to cut the head off a beating-heart cadaver and screw a different one onto it faces a significant hurdle in the form of donor consent. A single organ removed from a body becomes impersonal, identity-neutral. The humanitarian benefits of its donation outweigh the emotional discomfort surrounding its removal— for most of us, anyway. Body transplants are another story. Will people or their families ever give an entire, intact body away to improve the health of a stranger?

They might. It has happened before. Though these particular curative dead bodies never found their way to the operating room. They were more of an apothecary item: topically applied, distilled into a tincture, swallowed or eaten. Whole human bodies—as well as bits and pieces of them—were for centuries a mainstay in the pharmacopoeias of Europe and Asia. Some people actually volunteered for the job. If elderly men in twelfth-century Arabia were willing to donate themselves to become "human mummy confection" (see recipe, next chapter), then it's not hard to imagine that a man might volunteer to be someone else's transplanted body. Okay, it's maybe a little hard.

10
EAT ME

––––––

Medicinal cannibalism and the case
of the human dumplings

In the grand bazaars of twelfth-century Arabia, it was occasionally possible, if you knew where to look and you had a lot of cash and a tote bag you didn't care about, to procure an item known as mellified man. The verb "to mellify" comes from the Latin for honey, *mel*. Mellified man was dead human remains steeped in honey. Its other name was "human mummy confection," though this is misleading, for, unlike other honey-steeped Middle Eastern confections, this one did not get served for dessert. One administered it topically and, I am sorry to say, orally as medicine.

The preparation represented an extraordinary effort, both on the part of the confectioners and, more notably, on the part of the ingredients:

. . . In Arabia there are men 70 to 80 years old who are willing to give their bodies to save others. The subject does not eat food, he only bathes and partakes of honey. After a month he only excretes honey (the urine and feces are entirely honey) and death follows. His fellow men place him in a stone coffin full of honey in which he macerates. The date is put upon the coffin giving the year and month. After a hundred years the seals are removed. A confection is formed which is used for the treatment of broken and wounded limbs. A small amount taken internally will immediately cure the complaint.

The above recipe appears in the *Chinese Materia Medica,* a 1597 compendium of medicinal plants and animals compiled by the great naturalist Li Shih-chen. Li is careful to point out that he does not know for certain whether the mellified man story is true. This is less comforting than it sounds, for it means that when Li Shih-chen does *not* make a point of questioning the veracity of a *Materia Medica* entry, he feels certain that it is true. This tells us that the following were almost certainly used as medicine in sixteenth-century China: human dandruff ("best taken from a fat man"), human knee dirt, human ear wax, human perspiration, old drumskins ("ashed and applied to the penis for difficult urination"), "the juice squeezed out of pig's feces," and "dirt from the proximal end of a donkey's tail."

The medicinal use of mummified—though not usually mellified—humans is well documented in chemistry books of sixteenth-, seventeenth-, and eighteenth-century Europe, but nowhere outside Arabia were the corpses volunteers. The most sought-after mummies were said to be those of caravan members overcome by sandstorms in the Libyan desert. "This sudden suffocation doth concentrate the spirits in all the parts by reason of the fear and sudden surprisal which seizes on the travellers," wrote Nicolas Le Fèvre, author of *A Compleat Body of Chymistry.* (Sudden death also lessened the likelihood that the body was diseased.) Others claimed the mummy's medicinal properties derived from Dead Sea bitumen, a pitchlike substance which the Egyptians were thought, at the time, to have used as an embalming agent.

Needless to say, the real deal out of Libya was scarce. Le Fèvre offered a recipe for home-brewed mummy elixir using the remains of "a young, lusty man" (other writers further specified that the youth be a redhead). The requisite surprisal was to have been supplied by suffocation, hanging, or impalement. A recipe was provided for drying, smoking, and blending (one to three grains of mummy in a mixture of viper's flesh and spirit of wine) the flesh, but Le Fèvre offered no hint of how or where to procure it, short of suffocating or impaling the young carrot-top oneself.

There was for a time a trade in fake mummies being sold by Jews in Alexandria. They had apparently started out selling authentic mummies raided from crypts, prompting the author C. J. S. Thompson in *The Mystery and Art of the Apothecary* to observe that "the Jew eventually had his revenge on his ancient oppressors." When stocks of real mummies wore thin, the traders began concocting fakes. Pierre Pomet, private druggist to King Louis XIV, wrote in the 1737 edition of *A Compleat History of Druggs* that his colleague Guy de la Fontaine had traveled to Alexandria to "have ocular demonstration of what he had heard so much of" and found, in one man's shop, all manner of diseased and decayed bodies being doctored with pitch, wrapped in bandages, and dried in ovens. So common was this black market trade that pharmaceutical authorities like Pomet offered tips for prospective mummy shoppers: "Choose what is of a fine shining black, not full of bones and dirt, of good smell and which being burnt does not stink of pitch." A. C. Wootton, in his 1910 *Chronicles of Pharmacy*, writes that celebrated French surgeon and author Ambroise Paré claimed ersatz mummy was being made right in Paris, from desiccated corpses stolen from the gibbets under cover of night. Paré hastened to add that he never prescribed it. From what I can tell he was in the minority. Pomet wrote that he stocked it in his apothecary (though he averred that "its greatest use is for catching fish"). C. J. S. Thompson, whose book was published in 1929, claimed that human mummy could still be found at that time in the drug-bazaars of the Near East.

Mummy elixir was a rather striking example of the cure being worse than the complaint. Though it was prescribed for conditions ranging from palsy to vertigo, by far its most common use was as a treatment for contusions and preventing coagulation of blood: People were swallowing decayed human cadaver for the treatment of *bruises*. Seventeenth-century druggist Johann Becher, quoted in Wootton, maintained that it was "very beneficial in flatulency" (which, if he meant as a causative agent, I do not doubt). Other examples of human-sourced pharmaceuticals surely causing more distress than they relieved include strips of cadaver skin tied around the calves to prevent cramping, "old liquified placenta" to "quieten a patient whose hair stands up without cause" (I'm quoting Li Shih-chen on this one and the next), "clear liquid feces" for worms ("the smell will induce insects to crawl out of any of the body orifices and relieve irritation"), fresh blood injected into the face for eczema (popular in France at the time Thompson was writing), gallstone for hiccoughs, tartar of human teeth for wasp bite, tincture of human navel for sore throat, and the spittle of a woman applied to the eyes for ophthalmia. (The ancient Romans, Jews, and Chinese were all saliva enthusiasts, though as far as I can tell you couldn't use your own. Treatments would specify the type of spittle required: woman spittle, newborn man-child spittle, even Imperial Saliva, Roman emperors apparently contributing to a community spittoon for the welfare of the people. Most physicians delivered the substance by eyedropper, or prescribed it as a sort of tincture, although in Li Shih-chen's day, for cases of "nightmare due to attack by devils," the unfortunate sufferer was treated by "quietly spitting into the face.")

Even in cases of serious illness, the patient was sometimes better off ignoring the doctor's prescription. According to the

Chinese Materia Medica, diabetics were to be treated with "a cupful of urine from a public latrine." (Anticipating resistance, the text instructs that the heinous drink be "given secretly.") Another example comes from Nicholas Lemery, chemist and member of the Royal Academy of Sciences, who wrote that anthrax and plague could be treated with human excrement. Lemery did not take credit for the discovery, citing instead, in his *A Course of Chymistry,* a German named Homberg who in 1710 delivered *before the Royal Academy* a talk on the method of extracting "an admirable phosphorus from a man's excrements, which he found out after much application and pains"; Lemery reported the method in his book ("Take four ounces of humane Excrement newly made, of ordinary consistency . . ."). Homberg's fecal phosphorus was said to actually glow, an ocular demonstration of which I would give my eyeteeth (useful for the treatment for malaria, breast abscess, and eruptive smallpox) to see. Homberg may have been the first to make it glow, but he wasn't the first to prescribe it. The medical use of human feces had been around since Pliny's day. The *Chinese Materia Medica* prescribes it not only in liquid, ash, and soup form—for everything from epidemic fevers to the treatment of children's genital sores—but also in a "roasted" version. The thinking went that dung is essentially, in the case of the human variety,* bread and meat reduced to their simplest elements and thereby "rendered fit for the exercise of their virtues," to quote A. C. Wootton.

Not all cadaveric medicines were sold by professional druggists. The Colosseum featured occasional backstage concessions of blood from freshly slain gladiators, which was thought

---

* As opposed to the mouse, horse, rat, goose, hog, sheep, mule, donkey, or dog variety. Dog turd was especially popular, particularly dried white dog turd, from which the popular Renaissance medicine Album Graecum was made. The *Chinese Materia Medica* includes not only dog turd, but the grains and bones extracted from it. These were trying times for pharmacists.

to cure epilepsy,* but only if taken before it had cooled. In eighteenth-century Germany and France, executioners padded their pockets by collecting the blood that flowed from the necks of guillotined criminals; by this time blood was being prescribed not only for epilepsy, but for gout and dropsy.† As with mummy elixir, it was believed that for human blood to be curative it must come from a man who had died in a state of youth and vitality, not someone who had wasted away from disease; executed criminals fit the bill nicely. It was when the prescription called for bathing in the blood of infants, or the blood of virgins, that things began to turn ugly. The disease in question was most often leprosy, and the dosage was measured out in bathtubs rather than eyedroppers. When leprosy fell upon the princes of Egypt, wrote Pliny, "woe to the people, for in the bathing chambers, tubs were prepared, with human blood for the cure of it."

Often the executioners' stock included human fat as well, which was used to treat rheumatism, joint pain, and the poetic-sounding though probably quite painful falling-away limbs. Body snatchers were also said to ply the fat trade, as were sixteenth-century Dutch army surgeons in the war for independence from Spain, who used to rush onto the field

---

* If you could at all help it, it was extremely advisable, historically, to avoid being epileptic. Treatments for it have included distilled human skull, dried human heart, bolus of human mummy, boy's urine, excrement of mouse, goose, and horse, warm gladiator blood, arsenic, strychnine, cod liver oil, and borax.

† While I am thankful to be alive in the era of antibiotics and over-the-counter Gyne-Lotrimin, I am saddened by modern medicine's contributions to medical nomenclature. Where once we had scrofula and dropsy, now we have supraventricular tachyarrhythmia and glossopharyngeal neuralgia. Gone are quinsy, glanders, and farcy. So long, exuberant granulations and cerebral softening. Fare-thee-well, tetter and hectic fever. Even the treatments used to have an evocative, literary flavor. The *Merck Manual* of 1899 listed "a tumbler-ful of Carlsbad waters, sipped hot while dressing" as a remedy for constipation and the lovely, if enigmatic, "removal inland" as a cure for insomnia.

with their scalpels and buckets in the aftermath of a pitched battle. To compete with the bargain basement prices of the executioners, whose product was packaged and sold more or less like suet, seventeenth-century druggists would fancy up the goods by adding aromatic herbs and lyrical product names; seventeenth-century editions of the *Cordic Dispensatory* included Woman Butter and Poor Sinner's Fat. This had long been the practice with many of the druggists' less savory offerings: Druggists in the Middle Ages sold menstrual blood as Maid's Zenith and prettied it up with rosewater. C. J. S. Thompson's book includes a recipe for Spirit of the Brain of Man, which includes not only brain ("with all its membranes, arteries, veins and nerves"), but peony, black cherries, lavender, and lily.

Thompson writes that the rationale behind many of the human remedies was simple association. Turning yellow from jaundice? Try a glass of urine. Losing your hair? Rub your scalp with distilled hair elixir. Not right in the head? Have a snort of Spirit of Skull. Marrow and oil distilled from human bones were prescribed for rheumatism, and human urinary sediment was said to counteract bladder stones.

In some cases, unseemly human cures were grounded in a sort of sideways medical truth. Bile didn't cure deafness per se, but if your hearing problem was caused by a buildup of earwax, the acidy substance probably worked to dissolve it. Human toenail isn't a true emetic, but one can imagine that an oral dose might encourage vomiting. Likewise, "clear liquid feces" isn't a true antidote to poisonous mushrooms, but if getting mushrooms up and out of your patient's stomach is the aim, there's probably nothing quite as effective. The repellent nature of feces also explains its use as a topical application for prolapsed uterus. Since back before Hippocrates' day, physicians had viewed the female reproductive system not as an organ but as an independent entity, a mysterious creature with a will of its own, prone to haphazard "wanderings." If the uterus dropped down out of place following childbirth, a smear of something foul-smelling—often dung—was prescribed to coax it back up where it belonged. The active ingredient in human saliva was no doubt the natural antibiotic it contains; this would explain its use in treating dog bite, eye infection, and "fetid perspiration," even though no one at the time understood the mechanism.

Given that minor ailments such as bruises, coughs, dyspepsia, and flatulence disappear on their own in a matter of days, it's easy to see how rumors of efficacy came about. Controlled trials were unheard of; everything was based on anecdotal evidence. *We gave Mrs. Peterson some shit for her quinsy and now she's doing fine!* I talked to Robert Berkow, editor of the *Merck Manual*, for 104 years the best-selling physicians' reference book, about the genesis of bizarre and wholly unproven medicines. "When you consider that a sugar pill for pain relief will get a twenty-five to forty percent response," he said, "you can begin to understand how some of these treatments came to be recommended." It wasn't until about 1920, he added, that "the average patient with the average illness seeing the average physician came off better for the encounter."

The popularity of some of these human elixirs probably had less to do with the purported effective ingredient than with the base. The recipe in Thompson's book for a batch of King Charles' Drops—King Charles II ran a brisk side business in human skull tinctures out of his private laboratory in Whitehall—contained not only Spirit of Skull but a half pound of opium and four fingers (the unit of measurement, not the actual digits) of spirit of wine. Mouse, goose, and horse excrements, used by Europeans to treat epilepsy, were dissolved in wine or beer. Likewise powdered human penis, as prescribed in the *Chinese Materia Medica*, was "taken with alcohol." The stuff might not cure you, but it would ease the pain and put a shine on your mood.

Off-putting as cadaveric medicine may be, it is—like cultural differences in cuisine—mainly a matter of what you're accustomed to. Treating rheumatism with bone marrow or scrofula with sweat is scarcely more radical or ghoulish than treating, say, dwarfism with human growth hormone. We see nothing distasteful in injections of human blood, yet the thought of soaking in it makes us cringe. I'm not advocating a return to medicinal earwax, but a little calm is in order. As Bernard E. Read, editor of the 1976 edition of the *Chinese Materia Medica*, pointed out, "Today people are feverishly examining every type of animal tissue for active principles, hormones, vitamines and specific remedies for disease, and the discovery of adrenaline, insulin, theelin, menotoxin, and others, compels an open mind that one may reach beyond the unaesthetic setting of the subject to things worth while."

Those of us who undertook the experiment pooled our money to purchase cadavers from the city morgue, choosing the bodies of persons who had died of violence—who had been freshly killed and were not diseased or senile. We lived on this cannibal diet for two months and everyone's health improved.

So wrote the painter Diego Rivera in his memoir, *My Art, My Life*. He explains that he'd heard a story of a Parisian fur dealer who fed his cats cat meat to make their pelts firmer and glossier. And that in 1904, he and some fellow anatomy students—anatomy being a common requirement for art students—decided to try it for themselves. It's possible Rivera made this up, but it makes a lively introduction to modern-day human medicinals, so I thought I'd throw it in.

Outside of Rivera, the closest anyone has gotten to Spirit of Skull or Maid's Zenith in the twentieth century is in the medicinal use of cadaver blood. In 1928, a Soviet surgeon by the name of V. N. Shamov attempted to see if blood from the dead could be used in place of blood from live donors for transfusions. In the Soviet tradition, Shamov experimented first on dogs. Provided the blood was removed from the corpse within six hours, he found, the transfused canines showed no adverse reactions. For six to eight hours, the blood inside a dead body remains sterile and the red blood cells retain their oxygen-carrying capabilities.

Two years later, the Sklifosovsky Institute in Moscow got wind of Shamov's work and began trying it out on humans. So enamored of the technique were they that a special operating room was built to which cadavers were delivered. "The cadavers are brought by first-aid ambulances from the street, offices, and other places where sudden death overtakes human beings," wrote B. A. Petrov in the October 1959 issue of *Surgery*. Robert White, the neurosurgeon from Chapter 9, told me that during the Soviet era, cadavers belonged officially to the state, and if the state wanted to do something with them, then do something it did. (Presumably the bodies, once drained, were returned to the family.)

Corpses donate blood much the way people do, except that the needle goes in at the neck instead of the arm, and the body, lacking a working heart, has to be tilted so the blood pours out, rather than being pumped. The cadaver, wrote Petrov, was to be placed in "the extreme Trendelenburg position." His paper includes a line drawing of the jugular vein being entubed and a photograph of the special sterile ampules into which the blood flows, though in my opinion the space would have been better used to illustrate the intriguing and mysterious Trendelenburg position. I am intrigued only because I spent a month with a black-and-white photograph of the

"Sims position for gynecological examination"* on my wall, courtesy of the 2001 Mütter Museum calendar. ("The patient is to lie on the left side," wrote Dr. Sims. "The thighs are to be flexed, . . . the right being drawn up a little more than the left. The left arm is thrown behind across the back and the chest rotated forwards." It is a languorous, highly provocative position, and one has to wonder whether it was the ease of access it afforded or the similarity to cheesecake poses of the day that led our Dr. Sims to promote its use.)

The Trendelenburg position, I found out (by reading "Beyond the Trendelenburg Position: Friedrich Trendelenburg's Life and Surgical Contributions" in the journal *Surgery,* for I am easily distracted) simply refers to lying in a 45-degree incline; Trendelenburg used it during genitourinary surgery to tilt the abdominal organs up and out of the way. The paper's authors describe Trendelenburg as a great innovator, a giant in the field of surgery, and they mourn the fact that such an accomplished man is remembered for one of his slightest contributions to medical science. I will compound the crime by mentioning another of his slight contributions to medical science, the use of "Havana cigars to improve the foul hospital air." Ironically, the paper identified Trendelenburg as an outspoken critic of therapeutic bloodletting, though he registered no opinion on the cadaveric variety.

For twenty-eight years, the Sklifosovsky Institute happily transfused cadaver blood, some twenty-five tons of the stuff, meeting 70 percent of its clinics' needs. Oddly or not so oddly, cadaver blood donation failed to catch on outside the Soviet Union. In the United States, one man and one man alone

---

* You don't see the Sims position anymore, but you can see Dr. Sims, who lives on as a statue in Central Park in New York. If you don't believe me, you can look it up yourself, on page 56 of *The Romance of Proctology.* (Sims was apparently something of a dilettante when it came to bodily orifices.) P.S.: I could not, from cursory skimming, ascertain what the romance was.

dared try it. It seems Dr. Death earned his nickname long before it was given to him. In 1961, Jack Kevorkian drained four cadavers according to the Soviet protocol and transfused their blood into four living patients. All responded more or less as they would have had the donor been alive. Kevorkian did not tell the families of the dead blood donors what he was doing, using the rationale that blood is drained from bodies anyway during embalming. He also remained mum on the recipient end, opting not to tell his four unwitting subjects that the blood flowing into their veins came from a corpse. His rationale in this case was that the technique, having been done for thirty years in the Soviet Union, was clearly safe and that any objections the patients might have had would have been no more than an "emotional reaction to a new and slightly distasteful idea." It's the sort of defense that might work well for those maladjusted cooks that you hear about who delight in jerking off into the pasta sauce.

Of all the human parts and pieces mentioned in the *Chinese Materia Medica* and in the writings of Thompson, Lemery, and Pomet, I could find only one other in use as medicine today. Placenta is occasionally consumed by European and American women to stave off postpartum depression. You don't get placenta from the druggist as you did in Lemery's or Li Shih-chen's time (to relieve delirium, weakness, loss of willpower, and pinkeye); you cook and eat your own. The tradition is sufficiently mainstream to appear on a half-dozen pregnancy Web sites. The Virtual Birth Center tells us how to prepare Placenta Cocktail (8 oz. V-8, 2 ice cubes, 1/2 cup carrot, and 1/4 cup raw placenta, puréed in a blender for 10 seconds), Placenta Lasagna, and Placenta Pizza. The latter two suggest that someone other than Mom will be partaking—that it's being cooked up for dinner, say, or the PTA potluck—and one dearly hopes that the guests have been given a heads-up. The U.K.-based Mothers 35 Plus site lists "several sumptuous recipes," including roast placenta and dehydrated placenta. Ever the trailblazers, British television aired a garlic-fried placenta segment on the popular Channel 4 cooking show *TV Dinners*. Despite what one news report described as "sensitive" treatment of the subject, the segment, which ran in 1998, garnered nine viewer complaints and a slap on the wrist from the Broadcasting Standards Commission.

To see whether any of the human *Chinese Materia Medica* preparations are still used in modern China, I contacted the scholar and author Key Ray Chong, author of *Cannibalism in China*. Under the bland and benign-sounding heading "Medical Treatment for Loved Ones," Chong describes a rather gruesome historical phenomenon wherein children, most often daughters-in-law, were obliged to demonstrate filial piety to ailing parents, most often mothers-in-law, by hacking off a piece of themselves and preparing it as a restorative elixir. The practice began in earnest during the Sung Dynasty (960–1126) and continued through the Ming Dynasty, and up to the early 1900s. Chong presents the evidence in the form of a list, each entry detailing the source of the information, the donor, the beneficiary, the body part removed, and the type of dish prepared from it. Soups and porridges, always popular among the sick, were the most common dishes, though in two instances broiled flesh—one right breast and one thigh/upper arm combo—was served. In what may well be the earliest documented case of stomach reduction, one enterprising son presented his father with "lard of left waist." Though the list format is easy on the eyes, there are instances where one aches for more information: Did the young girl who gave her mother-in-law her left eyeball do so to prove the depth of her devotion, or to horrify and spite the woman? Examples for the Ming Dynasty were so numerous that Chong gave up on listing individual instances and presented them instead as tallies by category: In total, some 286 pieces of thigh, thirty-seven pieces of arm, twenty-four livers, thirteen unspecified cuts of flesh, four fingers, two ears, two broiled breasts, two ribs, one waist loin, one knee, and one stomach skin were fed to sickly elders.

Interestingly, Li Shih-chen disapproved of the practice. "Li Shih-chen acknowledged these practices among the ignorant masses," wrote Read, "but he did not consider that any parent, however ill, should expect such sacrifices from their children." Modern Chinese no doubt agree with him, though reports of the practice occasionally crop up. Chong cites a *Taiwan News* story from May 1987 in which a daughter cut off a piece of her thigh to cook up a cure for her ailing mother.

Although Chong writes in his book that "even today, in the People's Republic of China, the use of human fingers, toes, nails, dried urine, feces and breast milk are strongly recommended by the government to cure certain diseases" (he cites the 1977 *Chung Yao Ta Tz'u Tien,* the *Great Dictionary of Chinese Pharmacology*), he could not put me in touch with anyone who actually partakes, and I more or less abandoned my search. Then, several weeks later, an e-mail arrived from him. It contained a story from the *Japan Times* that week, entitled "Three Million Chinese Drink Urine." Around that same time, I happened upon a story on the Internet, originally published in the *London Daily Telegraph*, which based its story on one from the day before in the now-defunct *Hong Kong Eastern Express*. The article stated that private and state-run clinics and hospitals in Shenzhen, outside Hong Kong, sold or gave away aborted fetuses as a treatment for skin problems and asthma and as a general health tonic. "There are ten foetuses here, all aborted this morning," the *Express* reporter claims she was told while visiting the Shenzhen Health Centre for Women and Children undercover and asking for fetuses. "Normally we doctors take them home to eat. Since you don't look well, you can take them." The article bordered on the farcical. It had hospital cleaning women "fighting each other to take the treasured human remains home," sleazy unnamed chaps in Hong Kong back alleys charging $300 per fetus, and a sheepish businessman "introduced to foetuses by friends" furtively making his way to Shenzhen with his Thermos flask every couple of weeks to bring back "20 or 30 at a time" for his asthma.

In both this instance and that of the three million urine-quaffing Chinese, I didn't know whether the reports were true, partially true, or instances of bald-faced Chinese-bashing. Aiming to find out, I contacted Sandy Wan, a Chinese interpreter and researcher who had done work for me before in China. As it turned out, Sandy used to live in Shenzhen, had heard of the clinics mentioned in the article, and still had friends there—friends who were willing to pose, bless their hearts, as fetus-seeking patients. Her friends, a Miss Wu and a Mr. Gai, started out at the private clinics, saying they'd heard it was possible to buy fetuses for medicinal purposes. Both got the same answer: It used to be possible, but the government of Shenzhen had some time ago declared it illegal to sell both fetuses and placentas. The two were told that the materials were collected by a "health care production company with a unified management." It soon became clear what that meant and what was being done with the "materials." At the state-run Shenzhen People's Hospital, the region's largest, Miss Wu went to the Chinese medicine department to ask a doctor for treatment for the blemishes on her face. The doctor recommended a medication called Tai Bao Capsules, which were sold in the hospital dispensary for about $2.50 a bottle. When Miss Wu asked what the medication was, the doctor replied that it was made from abortus, as it is called there, and placenta, and that it was very good for the skin. Meanwhile, over in the internal medicine department, Mr. Gai had claimed to have asthma and told the doctor that his friends had recommended abortus. The doctor said he hadn't heard of selling fetuses to patients directly, and that they were taken away by a company controlled by the Board of Health, which was authorized to make them into capsules—the Tai Bao Capsules that had been prescribed to Miss Wu.

Sandy read the *Express* article to a friend who works as a doctor in Haikou, where the two women live. While her friend felt that the article was exaggerated, she also felt that fetal tissue did have health benefits and approved of making use of it. "It is a pity," she said, "to throw them away with other rubbish." (Sandy herself, a Christian, finds the practice immoral.)

It seems to me that the Chinese, relative to Americans, have a vastly more practical, less emotional outlook when it comes to what people put in their mouths. Tai Bao capsules notwithstanding, I'm with the Chinese. The fact that Americans love dogs doesn't make it immoral for the Chinese of Peixian city, who apparently don't love dogs, to wrap dog meat in pita bread and eat it for breakfast, just as the Hindu's reverence for cows doesn't make it wrong for us to make them into belts and meat loaves. We are all products of our upbringing, our culture, our need to conform. There are those (okay, one person) who feel that cannibalism has its place in a strictly rational society: "When man evolves a civilization higher than the mechanized but still primitive one he has now," wrote Diego Rivera in his memoir, "the eating of human flesh will be sanctioned. For then man will have thrown off all of his superstitions and irrational taboos."

Of course, the issue of taking fetus pills is complicated by the involvement and rights of the mother. If a hospital wants to sell—or even give away—women's aborted fetuses to make them into pills, they owe it to those women to ask for their consent. To do elsewise is callous and disrespectful.

Any attempt to market Tai Bao Capsules in the United States would be disastrous, owing to conservative religious views about the status of all fetuses as full-fledged human beings with all the rights and powers accorded their more cellularly differentiated brethren, and to good old-fashioned American squeamishness. The Chinese are simply not a squeamish people. Sandy once told me about a famous Chinese recipe called Scream Three Times, in which newborn mice are taken from their mothers (the first scream), dropped in a hot fry pot (second scream), and eaten (third scream). Then again, we drop live lobsters into boiling water and rid our homes of mice by gluing down their feet and letting them starve, so let us not rush to cast the first stone.

I began to wonder: Would any culture go so far as to use human flesh as food simply out of practicality?

China has a long and vivid history of cannibalism, but I'm not convinced that the taboo against it is any weaker there than elsewhere. Of the thousands of instances of cannibalism throughout China's history, the vast majority of the perpetrators were driven to the act either by starvation or the desire to express hatred or exact revenge during war. Indeed, without a strong cannibalism taboo, the eating of one's enemy's heart or liver would not have been the act of psychological brutality that it clearly was.

Key Ray Chong managed to unearth only ten cases of what he calls "taste cannibalism": eating the flesh or organs of the dead not because you have nothing else to eat or you despise your enemy or you're trying to cure an ailing parent, but simply because it's tasty and a pity to waste it. He writes that in years past, another job perk of the Chinese executioner—in addition to supplemental income from human blood and fat sales—was that he was allowed to take the heart and brains home for supper. In modern times, human meat for private consumption tends to come from murder victims—cannibalism providing at once a memorable repast and a handy means of disposing of the body. Chong relates the tale of a couple in Beijing who killed a teenager, cooked his flesh, and shared it with the neighbors, telling them it was camel meat. According to the story, which ran in the *Chinese Daily News* on April 8, 1985, the couple confessed that their motive had been a strong craving for human flesh, developed during wartime, when food was scarce. Chong doesn't find the story far-fetched. Because starvation cannibalism was widespread in the country's history, he believes that some Chinese, in certain hard-hit regions, over time may have developed a taste for human flesh.

It is said to be quite good. The Colorado prospector Alfred Packer, who, after his provisions ran out, began lunching on the five companions he was later accused of killing, told a reporter in 1883 that the breasts of men were "the sweetest meat" he'd ever tasted. A sailor on the damaged and drifting schooner *Sallie M. Steelman* in 1878 described the flesh of one of the dead crewman as being "as good as any beefsteak" he ever ate. Rivera—if we are to believe his anatomy lab tale—considered the legs, breasts, and breaded ribs of the female cadavers "delicacies," and especially relished "women's brains in vinaigrette."

Despite Chong's theory about Chinese people's occasionally acquiring a taste for human meat and despite China's natural culinary inhibition, instances of modern-day taste cannibalism are hard to find and even harder to verify. According to a 1991 Reuters article ("Diners Loved Human-Flesh Dumplings"), a man who worked in a crematorium in Hainan Province was caught hacking the buttocks and thighs off cadavers prior to incineration and bringing the meat to his brother, who ran the nearby White Temple Restaurant. For three years, the story went, Wang Guang was doing a brisk business in "Sichuan-style dumplings" made with flesh from the nether regions of his brother Hui's customers. The brothers were caught when the parents of a young woman killed in a road accident wanted to have a last look at her before cremation. "On discovering that her buttocks had been removed," wrote the reporter, "they

called the police." A second Reuters story on cannibalistic crematorium workers cropped up on May 6, 2002. The article
detailed the escapades of two Phnom Penh men accused—but
not prosecuted, for there was no law against cannibalism—of
eating human fingers and toes "washed down with wine."

The stories smacked of urban myth. Sandy Wan told me
she'd heard a similar story about a Chinese restaurant owner
who sees an accident and rushes over to slice off the buttocks
of the dead driver to use them as filling in steamed meat buns.
And the Hainan Reuters article had questionable elements:
How would the parents have seen their daughter's buttocks?
Presumably she was on her back in a coffin when they brought
her out for a final viewing. And why would the original article, from the *Hainan Special Zone Daily*, supply the names of the
men but not their town? Then again, this was Reuters. They
don't make things up. Do they?

Supper on China South Airways was an unsliced hamburger
bun and a puckered and unadorned wiener, rolling loose in a
pressed aluminum container. The wiener was too small for the
bun, too small for any bun, too small for its own skin. Even
for airline food, the meal was repugnant. The flight attendant,
having handed out the last of the meals, immediately about-
faced, returned to the front of the plane, and began picking
them up and dropping them into a garbage bag, on the just and
accurate assumption that no one was going to eat them.

If the White Temple Restaurant still existed, I would be able
to order an equally off-putting meal in about an hour. The
plane was landing shortly on Hainan Island, alleged home of
the buttock boys. I had been in Hong Kong and decided to
hop over to Hainan to look into the story. Hainan Province
turns out to be relatively small; it's an island off China's south-
west coast. The island has only one large city, Haikou, and

Haikou, I found out by e-mailing the Webmaster of the official Hainan Window Web site and pretending to be a funeral professional (a journalistic inquiry had gone unanswered), has a crematorium. If the story was true, this had to be where it happened. I would go to the crematorium and try to track down Hui and Wang Guang. I would ask them about their motives. Were they cheap and greedy, or were they simply practical—two well-meaning fellows who hated to see good meat go to waste? Did they see no wrong in their actions? Did they themselves eat and enjoy the dumplings? Did they think all human cadavers should be recycled this way?

My communications with the Hainan Webmaster had led me to believe that Haikou was a small, compact city, almost more of a town, and that most people spoke some English. The Web man did not have the address of the crematorium, but thought I could find it by asking around. "Even just ask a taxi driver," he wrote.

It took a half hour to even just ask a taxi driver to take me to my hotel. Like all taxi drivers and almost everyone else in Haikou, he spoke no English. Why should he? Few foreigners come to Hainan, only holiday-making Chinese from the mainland. The driver eventually telephoned a friend who spoke some English and I found myself in the midst of a vast, urban sprawl in a modern high-rise with huge red Chinese characters on its roof spelling out, I supposed, the hotel's name. Chinese big-city hotel rooms are modeled after their Western counterparts, with triangulated toilet paper ends and complimentary shower caps; however, there is always something slightly, ever so charmingly off. Here, it was a tiny bottle labeled "Sham Poo" and a flyer offering the services of a blind masseuse. (*Oh, madam! I'm so sorry! I thought that was your back! You see I'm blind. . . .*) Exhausted, I collapsed on the bed, which made a shrieking, assaulted noise, suggesting that it could as easily have been the bed that collapsed on me.

In the morning I approached the reception desk. One of the girls spoke a little English, which was helpful, though she had an unsettling habit of saying "Are you okay?" in place of "How are you?" as though I'd tripped on the rug coming out of the elevator. She understood "taxi" and pointed to one outside.

The night before, in preparation for my journey, I had drawn a picture to give to the cabdriver. It showed a body hovering above flames, and to the right of this I drew an urn, though the latter had come out looking like a samovar, and there was a distinct possibility that the driver would think I was looking for a place to get Mongolian barbecue. The driver looked at the piece of paper, seemed to understand, and pulled out into traffic. We drove for a long time, and it seemed we might actually be headed for the outskirts of town, where the crematorium was said to be. And then I saw my hotel go by on the right. We were driving in circles. What was going on? Did the blind masseuse moonlight as a cabdriver? This was not good. I was not okay. I motioned to my merrily revolving driver to pull over, and I pointed to the Chinese Tourism Agency office on the map.

Eventually the taxi pulled up outside a brightly lit fried chicken establishment, the sort of place that in the United States might proclaim "We Do Chicken Right!" but here proclaimed "Do Me Chicken!" The cabdriver turned to collect his fare. We shouted at each other for a while, and eventually he got out and walked over to a tiny, dim storefront next to the chicken place and pointed vigorously to a sign. Designated Foreign-Oriented Tourist Unit, it said. Well, do me chicken. The man was right.

Inside, the tourist unit was having a cigarette break, which, judging from the density of the smoke, had been going on for some time, years possibly. The walls were bare cement and part of the ceiling was falling in. There were no travel brochures or train timetables, only a map of the world and a small wall-mounted shrine with a red electric candle and a bowl of offerings. The gods were having apples. In the back of the office, I could see two brand-new shrink-wrapped chairs. This struck me as an odd purchasing decision, what with the ceiling collapsing and the very slim likelihood that more than two or three tourists a year came in and needed a place to sit.

I explained to the woman that I wanted to hire an interpreter. Miraculously, two phone calls and half an hour later, one appeared. It was Sandy Wan, the woman who would later help me track down the truth about the abortus vendors. I explained that I needed to talk to someone at the Haikou crematorium. Sandy's English vocabulary was impressive but, understandably, did not include "crematorium."

I described it as the big building where they burn dead bodies. She didn't catch the last bit and thought I meant some sort of factory. "What kind of material?" she asked. The entire staff of the designated foreign-oriented tourist unit were looking on, trying to follow the conversation.

"Dead people . . . material." I smiled helplessly. "Dead bodies."

"Ah," said Sandy. She did not flinch. She explained to the tourist unit, who nodded as though they got this sort of thing all the time. Then she asked me for the address. When I replied that I didn't know it, she got the crematorium phone number from the information operator, called the place to get the address, and even set up an appointment with the director. She was amazing. I couldn't imagine what she had told the man, or what she thought I needed to talk to him about. I began to feel a little sorry for the crematorium director, thinking he was about to be visited by a grieving foreign widow, or perhaps some glad-handing retort salesman there to help him cut costs and maximize efficiency.

In the cab, I tried to think of a way to explain to Sandy what I was about to have her do. *I need you to ask this man whether he*

had an employee who cut the butt cheeks off cadavers to serve in his
brother's restaurant. No matter how I thought of phrasing it, it
sounded ghastly and absurd. Why would I need to know this?
What kind of book was I writing? Fearing that Sandy might
change her mind, I said nothing about the dumplings. I said
that I was writing an article for a funeral industry magazine.
We were outside the city proper now. Trucks and scooters
had gone scarce. People drove wooden ox carts and wore the
round, peaked sun hats you see in rural Vietnam, only these
were fashioned from laminated newspaper. I wondered if
someone, somewhere, was wearing the March 23, 1991, edition of the *Hainan Special Zone Daily*.

The taxi turned off onto a dirt road. We passed a brick
smokestack, issuing clouds of black: the crematorium. Farther
down the road was the accompanying funeral home and the
crematorium offices. We were directed up a broad marble
stairway to the director's office. This could only go poorly.
The Chinese are wary of reporters, especially foreign ones,
and very especially foreign ones suggesting that your staff
mutilated the dead relations of paying customers to make
dumplings. What had I been thinking?

The director's office was large and sparsely furnished. There
was nothing on the walls but a clock, as if no one knew how
to decorate for death. Sandy and I were seated in leather chairs
that sat low to the floor, like car seats, and told that the director would be in to see us shortly. Sandy smiled at me, unaware
of the horror about to unfold. "Sandy," I blurted out, "I have
to tell you what this is about! There was this guy who cut the
butts off dead bodies to give to his brother to . . ."

It was at that moment that the director walked in. The
director was a stern-looking Chinese woman, easily six feet
tall. From my humbled position near the floor, she seemed to
be of superhuman proportions, as tall as the smokestack outside
and as likely to belch forth smoke.

The director sat down at her desk. She looked at me. Sandy looked at me. Feeling seasick, I launched into my story. Sandy listened and, bless her, betrayed no emotion. She turned to the director, who was not smiling, had not smiled since she entered the room, had possibly never smiled, and she told her what I had just said. She relayed the story of Hui Guang, explained that I thought he might have been employed here, and that I wrote for a magazine and that I hoped to find him and speak to him. The director crossed her arms and her eyes narrowed. I thought I saw her nostrils flare. Her reply went on for ten minutes. Sandy nodded politely through it all, with the attentive calm of a person being given a fast-food order or directions to the mall. I was very impressed. Then she turned to me. "The director, she is, ah, very angry. The director is very . . . *astonished* to have these facts. She never heard of this story. She says she has known all her workers, and she has been here for more than ten years and she would know about this kind of story. Also, she feels it is a . . . really sick story. And so she cannot help you." I would love to see a full transcript of the director's reply, and then again I wouldn't.

Back in the cab I explained myself to Sandy as best I could. I apologized for putting her through this. She laughed. We both laughed. We laughed so hard that the cab driver demanded to know what we were laughing about, and he laughed too. The cab driver had grown up in Haikou, but he hadn't heard the story of the Guang brothers. Neither, it later turned out, had any of Sandy's friends. We had the driver let us off at the Haikou public library to look for the original article. As it turns out, no paper named the *Hainan Special Zone Daily* exists, only the *Hainan Special Zone Times,* which is a weekly. Sandy looked through the papers for the week of March 23, 1991, but there was no mention of the human dumplings. She also checked old phone books for the White Temple Restaurant and found nothing.

There wasn't much more to do in Haikou, so I took the bus
south to Sanya, where the beaches are beautiful and the weather
is fine and there is, I found out, another crematorium. (Sandy
called the director and received a similarly indignant reply.)
On the beach that afternoon, I spread my towel a few feet away
from a wooden sign that advised beach-goers, "Do not spit at
the beach." Unless, I thought to myself, the beach suffers from
nightmares, ulcers, ophthalmia, or fetid perspiration.

Anthropologists will tell you that the reason people never
dined regularly on other people is economics. While there
existed, I am told, cultures in Central America that actually
ranched humans—kept enemy soldiers captive for a while to
fatten them up—it was not practical to do so, because you had
to give up more food to feed them than you'd gain in the end
by eating them. Carnivores and omnivores, in other words,
make lousy livestock. "Humans are very inefficient in con-
verting calories into body composition," said Stanley Garn,
a retired anthropologist with the Center for Human Growth
and Development at the University of Michigan. I had called
him because he wrote an *American Anthropologist* paper on the
topic of human flesh and its nutritional value. "Your cows," he
said, "are much more efficient."

But I am not so much interested in cultures' eating the flesh
of their captive enemies as I am in cultures' eating their own
dead: the practical, why-not model of cannibalism—eating the
meat of fresh corpses because it's there and it's a nice change
from taro root. If you're not going out and capturing people
and/or going to the trouble of fattening them up, then the
nutritional economics begin to make more sense.

I found an *American Anthropologist* article—a reply to
Garn's—stating that there are in fact instances of groups of
humans who will eat not only enemies they have killed, but

members of their own group who have died of natural causes. Though in every case, the author, University of California, San Diego, anthropologist Stanley Walens, said, the cannibalism was couched in ritual. No culture, as far as he knew, simply carved up dead tribe members to distribute as meat.

Garn seemed to disagree. "Lots of cultures ate their dead," he said, though I couldn't get any specifics out of him. He added that many groups—too many, he said, to specify—would eat infants as a means of population control when food was scarce. Did they kill them or were they already dead, I wanted to know.

"Well," he replied, "they were dead by the time they ate them." This is how conversations with Stanley Garn seem to go. Somehow, midway through our chat, he steered the conversation from nutritional cannibalism to the history of landfill—a pretty sharp turn—and there it more or less remained. "You should write a book about *that*," he said, and I think he meant it.

I had called Stanley Garn because I was looking for an anthropologist who had done a nutritional analysis of human flesh and/or organ meats. Just, you know, curious. Garn hadn't exactly done this, but he had worked out the lean/fat percentage of human flesh. He estimates that humans have more or less the same body composition as veal. To arrive at the figure, Garn extrapolated from average human body fat percentages. "There's information of that sort on people in most countries now," he said. "So you can see who you want for dinner." I wondered how far the beef/human analogy carried. Was it true of human flesh, as of beef, that a cut with more fat is considered more flavorful? Yup, said Garn. And, as with livestock, the better nourished the individuals, the higher the protein content. "The little people of the world," said Garn—and I had to assume he was referring to the malnourished denizens of the third world and not dwarfs—"are hardly worth eating."

There is only one group of contemporary individuals whose daily diet ran the risk of containing their own dead, and that is the California pet. In 1989, while researching a story on a ridiculous and racist law aimed at preventing Asian immigrants from eating their neighbors' dogs (which was already illegal because it's illegal to steal a dog), I learned that, owing to California Clean Air Act regulations, humane societies had switched from cremating euthanized pets to what one official called "the rendering situation." I called up a rendering plant to learn into what the dogs were being rendered. "We grind 'em up and turn 'em into bonemeal," the plant manager had said. Bonemeal is a common ingredient in fertilizers and animal feed—including many commercial animal foods.

Happily, pet food manufacturers have stayed away from this particular cost-cutting solution. In 2001, the FDA's Center for Veterinary Medicine tested an array of commercial pet foods to see if they contained DNA of dogs or cats. None was found.

Of course, no humans are made into fertilizer after they're dead. Or not, anyway, unless they wish to be.

11

OUT OF THE FIRE,
INTO THE COMPOST BIN

———

And other new ways to end up

When a cow dies on a visit to the hospital, it does not go to a morgue. It goes to a walk-in refrigerator, such as the one at Colorado State University Veterinary Teaching Hospital, in Fort Collins. Like most things in walk-in refrigerators, the bodies here are arranged to maximize space. Against one wall, sheep are stacked like sandbags against a flood. Cows hang from ceiling hooks, effecting the familiar side-of-beef silhouette. A horse, bisected mid-torso, lies in halves on the floor, a vaudeville costume after the show.

The death of a farm animal is death reduced to the physical and the practical: a matter of waste disposal and little more. With no soul to be ushered onward, no mourners to attend to, death's overseers are free to pursue more practical approaches. Is there a more economical way to dispose of the body? A more environmentally friendly way? Could something useful be done with the remains? With our own deaths, the disposal of the body was for centuries incorporated into the ritual of memorial and farewell. Mourners are present at the lowering of the coffin and, until more recently, the measured, remote-control conveyance of the casket into the cremation furnace. With the majority of cremations now done out of view of the mourners, the memorial has begun to be separated from the process of disposal. Does this free us to explore new possibilities?

Kevin McCabe, owner of McCabe Funeral Homes in Farmington Hills, Michigan, is one man who thinks that the answer is yes. One day soon, he plans to do to dead people what Colorado State University is doing to dead sheep and horses. The process—called "tissue digestion" when you speak to the livestock people and "water reduction" when you speak to McCabe—was invented by a retired pathology professor named Gordon Kaye and a retired professor of biochemistry named Bruce Weber. McCabe is the mortuary consultant for Kaye and Weber's company, WR2 , Inc., based in Indianapolis, Indiana.

The mortuary end of corpse disposal had been a low priority over at WR2 until the spring of 2002, when Ray Brant Marsh of Noble, Georgia, dragged the good name of crematory operators everywhere about as far through the mud as a name could go. At last count, some 339 decomposing bodies were found on land surrounding his Tri-State Crematory—stacked in sheds, dumped in a pond, crammed in a concrete burial vault. Marsh initially claimed the incinerator wasn't working, but it was. Then rumors of decomposing body photos in his computer files made the rounds. It began to look as though Marsh wasn't simply cheap and unethical, but deeply strange. As the body count grew, Gordon Kaye began to get calls: half a dozen from funeral directors, and one from a New York State assemblyman, all wanting to know how soon the mortuary tissue digestor might be available, should the public begin to shun crematoriums. (At that time, Kaye estimated it would be another six months.)

In a few hours, Kaye and Weber's equipment can dissolve the tissues of a corpse and reduce it to 2 or 3 percent of its body weight. What remains is a pile of decollagenated bones that can be crumbled in one's fingers. Everything else has been turned into what the WR2 brochure describes as a sterile "coffee-colored" liquid.

Tissue digestion relies on two key ingredients: water and an alkali better known as lye. When you put lye into water, you create a pH environment that frees the hydrogen ion of the water to break apart the proteins and fats that make up a living organism. That's why "water reduction," though clearly a euphemism, is an apt term. "You are using water to break the chemical bonds in the large molecules of the body," says Kaye. But Kaye does not gloss over the lye. This is a man who has spent eleven years in the world of carcass disposal (or "disposition," if you are speaking with McCabe). "In effect, it's a pressure cooker with Drano," says Kaye of his invention. The lye does more or less what it would do if you swallowed it. You don't digest it, it digests you. What's nice about an alkali, as opposed to an acid, is that in doing the deed, the chemical renders itself inert and can be safely flushed down the drain.

There is no question that tissue digestion makes good sense for disposing of dead animals. It destroys pathogens, and, more important, it destroys prions—including the ones that cause mad cow disease—which rendering cannot reliably do. It does not pollute, as incinerators do. And because no natural gas is used, the process is approximately ten times cheaper than incineration.

What are the advantages for humans? If they're humans who own funeral homes, the advantage is economical. A mortuary digestor will be relatively inexpensive to buy (less than $100,000) and, as mentioned, a tenth as expensive to run. Digestors make especially good sense in rural areas whose populations are too small to keep a crematory furnace continuously active, which is the best way for it to be. (Firing it up and letting it cool all the way down and refiring it over and over damages the furnace lining; ideally, you want to keep the fire going nonstop, turning it down just low enough to remove the ashes and put the next body in, but this presumes a steady lineup of corpses.)

What are the advantages for humans who don't own funeral homes? Assuming it's going to cost a family more or less the same as cremation would, why would someone choose to have this done? I asked McCabe, a chatty, affable Midwesterner, how he plans to market the process to bereaved families. "Simple," he said. "To families who come in and say, 'I want him to be cremated,' I'm gonna say, 'No problem. You can cremate him, or you can do our water reduction process.' And they're gonna say, 'What's that?' And I'm gonna go, 'Well, it's like cremation, but we do it with water under pressure instead of fire.' And they're gonna go, 'All right! Let's do it!' "

And the media is gonna go, "There's lye in there. You're boiling them in lye!" I mean, Kevin, I said, aren't you leaving out a pretty big part of it? "Oh, yeah, they're gonna know all that," he said. "I've talked to people and they have no problem." I'm not sure I believe him on these two points, but I do believe what he said next: "Besides, watching somebody cremated is not pretty."

I decided I had to see the process for myself. I contacted the chairman of the state anatomical board in Gainesville, Florida, where for the past five years digestors have been taking care of anatomy lab leftovers—here under the name "reductive cremation," in order to hopscotch state regulations that willed bodies be cremated. When I got no reply, Kaye gave me a contact at Colorado State. And that is how I came to be standing in a walk-in refrigerator full of dead livestock in Fort Collins, Colorado.

The digestor sits on a loading dock, fifteen feet from the walk-in. It is a round stainless-steel vat similar in size and circumference to a California hot tub. Indeed, when full, the two hold approximately the same mass of heated liquid and passive bodies: about seventeen hundred pounds.

Manning the digestor this afternoon is a soft-voiced wildlife pathologist named Terry Spracher. Spracher wears rubber boots pulled over his pants, and latex gloves. Both are streaked with blood, for he has been doing sheep necropsies.* Despite what his job duties might suggest, this is a man who loves animals. When he heard I lived in San Francisco, he brightened and said that he enjoyed visiting the city, and the reason he enjoyed it was not the hills or the Wharf or the restaurants but the Marine Mammal Center, an obscure ecology center up the coast where oil-soaked otters and orphaned elephant seals are rehabbed and released. I guess this is how it is with animal careers. If you deal with animals for a living, you generally also deal with their deaths.

Above our heads, the unit's perforated liner basket hangs from a ceiling-mounted hydraulic hoist on a track. A taciturn, ginger-haired lab assistant named Wade Clemons pushes a button, and the basket travels across the loading dock to the door of the walk-in, where he is standing. When he's done loading the basket, he and Spracher will guide it back to the airspace above the digestor and lower it in. "Just like french fries," says Spracher quietly.

Hanging from the hoist inside the walk-in is a large steel hook. Clemons bends down to couple this to a second hook, anchored on a thick band of muscle at the base of the horse's neck. Clemons presses a button. The half-horse rises. The sight is a disquieting blend of horse-as-we-know-it—placid, dejected horse face; silken mane and neck where young girls' hands went—and slasher-flick gore.

Clemons loads one half, then the other, lowering it down

---

* He does not use the word "autopsy," for the prefix denotes a postmortem medical inspection of one's own species. Technically speaking, only a human's investigation of another human's death can be called an autopsy—or, supposing a very different world, a sheep's investigation of another sheep's.

in beside its partner, the two halves fitting neatly together like new shoes in a box. With the seasoned expertise of a grocery bagger, Clemons loads sheep, a calf, and the nameless slippery contents of two ninety-gallon "gut buckets" from the necropsy lab, until the basket is full.

Then he presses a button that sends the basket along the ceiling track on a short, slow trip across the loading dock to the digestor. I try to imagine a cluster of mourners standing by, as they have stood by gravesides as winches lower coffins, and in cremation parlors as coffins on conveyor belts are pulled slowly into crematory retorts. Of course, for mortuary digestions, some alterations will be made in the name of dignity. The mortuary model will use a cylindrical basket and will process only one body at a time. McCabe doesn't see this as something the family would stand around and watch, though "if they wanted to see the equipment, they'd be welcome."

With the basket in place, Spracher closes the digestor's steel hatch and presses a series of buttons on the computerized console. Washing-machine noises can be heard as water and chemicals pour into the tank.

I return for the raising of the basket, the following day. (The process normally takes six hours for a load this size, but Colorado State needs to upgrade its pipes.) Spracher unbolts the hatch and raises the lid. I don't smell anything, and am emboldened to lean my head over the vat and peer inside. Now I smell something. It is a large, assertive smell, unappetizing and unfamiliar. Gordon Kaye refers to the smell as "soaplike," leading one to wonder where he buys his toiletries. The basket appears largely empty, which is pretty amazing when you think about what it looked like going in. Clemons turns on the hoist, and the basket rises from the machine. At the bottom is a foot and a half of bone hulls. I resolve to take Kaye's word for it that you can crumble them in your fingers.

Clemons opens a small door near the base of the basket and scrapes the bones out into a Dumpster. Though it's no more grisly than the emptying of a crematory retort, it's hard for me to imagine this catching on as part of the American funerary tradition. But here again, the funerary rendition wouldn't go quite like this. Had this been a mortuary digestion, the bone remnants would be dried and either pulverized for scattering or, as McCabe envisions, placed in a "bone box," a sort of mini-coffin that could be stored in a crypt or buried.

Everything other than bone has liquefied and disappeared down the drain. When I got back home I asked McCabe how he was going to handle the potentially disturbing realities of the dearly departed's molecules ending up in the municipal sewer system. "The public seems okay with it," he said. Contrasting it with cremation, he said, "You're either going to go in the sewer or you're going to go up in the atmosphere. People who are environmentally conscious know that we're better off putting something sterile and pH-neutral into the sewer than we are letting mercury [from fillings] go into the air."[*] McCabe is counting on environmental conscience to sell the process. Will it work?

---

[*] In the grand scheme of industrial air pollution, crematoria rank low on the fret list. They emit about half as much particulate matter as a residential fireplace and about as much nitrous oxide as the typical restaurant grill. (This is not surprising, as the human body is mostly water.) Of greatest concern is mercury from dental fillings, which vaporizes and drifts into the atmosphere at a rate of .23 grams per hour of operation (about a half gram per cremation), according to research done jointly by the EPA and the Cremation Association of North America. An independent study done in England in 1990 and published in the journal *Nature* estimated the average amount of mercury released into the atmosphere at three grams per cremation—a notably higher and, the author believed, worrisome total. All in all, compared to power plants and incinerated trash, the dental work of the dead generates a small fraction of the planet's airborne mercury.

We'll soon see. McCabe is poised to take delivery of the world's first mortuary tissue digestor sometime in 2003.

You have only to look at the story of cremation to appreciate that changing the way America disposes of its dead is a feat not easily accomplished. The best way to do this would be to buy a copy of Stephen Prothero's *Purified by Fire: A History of Cremation in America.* Prothero is a professor of religion at Boston University, a masterful writer, and a respected historian; his book includes a bibliography of more than two hundred original and secondary sources. The second-best way to do it would be to read the passage that follows, which is basically small chunks of Prothero's book run through the tissue digestor of my brain.

Ironically, one of the cremationists' earliest and loudest arguments in America was that cremation was less polluting than burial. In the mid-1800s, it was widely (and wrongly) believed that buried, decomposing bodies gave off noxious gases which polluted the groundwater and made their way up through the dirt to form deadly, hovering graveyard "miasmas" that tainted the air and sickened those who wandered past. Cremation was presented as the pure and hygienic alternative and might well have caught on then, had the first U.S. cremation not proved to be a PR disaster.

America's first crematory was built in 1874, on the estate of Francis Julius LeMoyne, a retired physician, abolitionist, and champion of education. Though his credentials as a social reformer were impressive, his beliefs about personal hygiene may have worked against him in his crusade for funereal cleanliness and purity. According to Prothero, he believed that "the human body was never intended by its Creator to come in contact with water," and, as such, traveled about in his own personal miasma.

LeMoyne's first customer was one Baron Le Palm, who was to be incinerated in a public ceremony to which national and European press had been invited. Le Palm's reasons for requesting cremation remain murky, but somewhere in the mix was a deep-seated fear of live burial, for he claimed to have met a woman who had been buried alive (presumably not very deeply). As things turned out, Mr. Le Palm was finished some months before the crematory was, and had to be preserved. He fell victim to the spotty and improvisational embalming techniques of the day, and wasn't looking his best when rowdier elements of the crowd—uninvited townsfolk, mainly—pulled the sheet from his earthly remains. Crude jokes were made. Schoolchildren snickered. Reporters from newspapers across the country criticized the carnival air of the proceedings and the lack of religious ritual and due solemnity. Cremation was all but doomed to an early grave.

Prothero posits that LeMoyne had erred in presenting a more or less secular ceremony. His unsentimental memorial speech, devoid of references to the Hereafter and the Almighty, and the bare, utilitarian design of his crematory (reporters likened it to "a bake oven" and "a large cigar box") offended the sensibilities of Americans used to Victorian-style funerals with their formal masses and their profusely flowered, ornately appointed caskets. America was not ready for pagan funerals. It would not be until 1963—when the Catholic Church, in the wake of the reforms of Vatican II, relaxed the ban on cremation—that disposal by incineration would start to take hold in a serious way. (1963 was a banner year for cremation. It was that summer that *The American Way of Death*, the late Jessica Mitford's exposé of deceit and greed in the burying business, came out.)

What has inspired funeral reformers throughout history, Prothero maintains, has been a distaste for pomp and religious pageantry. They may hand out pamphlets detailing the horrors and health risks of the grave, but what really bothered them was the waste and fakery of the traditional Christian funeral: the rococo coffins, the hired mourners, the expense, the wasted land. Freethinkers like LeMoyne envisioned a purer, simpler, back-to-basics approach. Unfortunately, as Prothero points out, these men have tended to take mortuary utilitarianism too far, outraging the churches and alienating the public. Take the American doctor who put forth a plan to boost the dead's utility by skinning them prior to cremation and making leather. Take the Italian professor who advocated burning cadaveric fat in streetlamps, speculating that the 250 people who died each day in New York would yield 30,600 pounds of fuel daily. Take the cremationist Sir Henry Thompson, who sat down and calculated the value in pounds sterling of the 80,000-odd people who died each year in London, should their cremated remains be used as fertilizer. It worked out to about £50,000, though the customers, should any have emerged, would have been dealt a raw deal, as cremains make lousy fertilizer. If you wanted to fertilize your garden with dead people, you were better off doing it the Hay way. Dr. George Hay was a Pittsburgh chemist who advocated pulverizing dead bodies so that they would—to quote an 1888 newspaper article on the topic—"return to the elements as soon as possible, if for no other purpose than to furnish a fertilizer." Here is Hay, quoted at length in the article, which is pasted into a scrapbook belonging to the Historical Collection of the Mount Auburn Cemetery in Cambridge, Massachusetts:

> The machines might be so contrived as to break the bones first in pieces the size of a hen egg, next into fragments of the size of a marble, and the mangled and lacerated mass could next be reduced by means of chopping machines and steam power to mincemeat. At this stage we have a homogeneous mixture of the entire body structures in the form of a

pulpous mass of raw meat and raw bones. This mass should now be dried thoroughly by means of steam heat at a temperature of 250 Fahr. . . . because firstly we wish to reduce the material to a condition convenient for handling and secondly we wish to disinfect it. . . . Once in this condition, it would command a good price for the purpose of manure.

Which brings us, ready or not, to the modern human compost movement. Here we must travel to Sweden, to a tiny island called Lyrön, due west of Gothenburg. This is the home of a forty-seven-year-old biologist-entrepreneur named Susanne Wiigh-Masak. Two years ago, Wiigh-Masak founded a company called Promessa, which seeks to replace cremation (the choice of 70 percent of Swedes) with a technologically enhanced form of organic composting. This is no mom-and-pop undertaking of the lunatic Green fringe. Wiigh-Masak has King Carl Gustav and the Church of Sweden on her side. She has crematoria vying to be the first to compost a dead Swede. She has the dead Swede ready to go (a terminally ill man who contacted her after hearing her on the radio; he has since taken up residence in a freezer in Stockholm). She has major corporate backing, an international patent, over two hundred press clips. Mortuary professionals and entrepreneurs from Germany, Holland, Israel, Australia, and the United States have expressed interest in representing Promessa's technology in their own countries.

She appears to be doing, in a matter of years, what took the cremationists a century.

This is especially impressive given that what she is proposing has its closest precedent in the ideas of Dr. George Hay. Let's say a man dies in Upsala, and that he has checked the box on the church-distributed living will that says, "I want that the new method freeze-drying ecological funeral will be used if it is available when I die." (The equipment is still being developed; Wiigh-Masak hopes to have it ready sometime in 2003.) The man's body will be brought to an establishment that has licensed Promessa's technology. He will be lowered into a vat of liquid nitrogen and frozen. From here he will progress to the second chamber, where either ultrasound waves or mechanical vibration will be used to break his easily shattered self* into small pieces, more or less the size of ground chuck. The pieces, still frozen, will then be freeze-dried and used as compost for a memorial tree or shrub, either in a churchyard memorial park or in the family's yard.

The difference between George Hay and Susanne Wiigh-Masak is that Hay, in suggesting that we feed crops with the dead, was simply trying to be practical, to do something beneficial and useful with a dead human body. Wiigh-Masak is not a utilitarian. She is an environmentalist. And in parts of Europe, environmentalism is tantamount to its own religion. For this reason, I think, she may just succeed.

To understand Wiigh-Masak's catechism, it helps to pay a visit to her compost pile. It lies beside the barn on the acre and

---

* Frozen humans shatter easily because they are mostly water. How much water is a matter of some debate. A Google search unearthed sixty-four Web sites with the words "body is 70 percent water," 27 sites that say it's 60 percent water, 43 that tell you it is either 80 or 85 percent water, 12 that say the figure is 90 percent, 3 that say it's 98 percent, and one that says it's 91 percent. A better consensus exists for jellyfish. They are either 98 or 99 percent water, and that is why you seldom see dried jellyfish snacks.

Todd Astorino, director of the Exercise Science Program at Salisbury University, in Salisbury, Maryland, was able to answer the question not only with certainty, but to a decimal point: We are 73.8 percent water. The figure, he said, is calculated by giving a volunteer a measured quantity of water laced with tracers to drink. Four hours later, the subject's blood is sampled and the dilution of the tracers is noted. From this, you, or Todd anyway, can figure out how much water is in the body. (The more water in the body, the more diluted the tracers in the blood.) Compare the water weight to body weight, and there's the answer. Isn't science terrific?

a half that she and her family rent on Lyrön. Wiigh-Masak shows her compost pile to guests the way an American home-owner might show off the new entertainment center, or the youngest son's grades. It is her pride and, it is no exaggeration to say, her joy.

She pushes a shovel into the heap and raises a loamy clod. It is complex and full of unnamable fragments, like a lasagna baked by an unsupervised child. She points out feathers from a. duck that died a few weeks back, shells from the mussels that her husband, Peter, farms on the other side of the island, cabbage from last week's coleslaw. She explains the difference between rotting and composting, that the needs of humans and the needs of compost are similar: oxygen, water, air tem-perature that does not stray far from 37 degrees centigrade. Her point: We are all nature, all made of the same basic mate-rials, with the same basic needs. We are no different, on a very basic level, from the ducks and the mussels and last week's coleslaw. Thus we should respect Nature, and when we die, we should give ourselves back to the earth.

As though sensing that she and I might not be entirely on the same page, perhaps not even in the same general vicin-ity of the Dewey decimal system, Wiigh-Masak asks me if I compost. I explain that I don't have a garden. "Ah, okay." She considers this fact. I get the feeling that to Wiigh-Masak, this is not so much an explanation as a criminal confession. I am feeling more like last week's coleslaw than usual.

She returns to the clod. "Compost should not be ugly," she is saying. "It should be lovely, it should be romantic." She feels similarly about dead bodies. "Death is a possibility for new life. The body becomes something else. I would like that that something else be as positive as possible." People have criti-cized her, she says, for lowering the dead to the level of garden waste. She doesn't see it that way. "I say, let's lift garden waste to as high a level as human bodies." What's she's trying to say is that nothing organic should be treated as waste. It should all be recycled.

I am waiting for Wiigh-Masak to put down the shovel, but now it is coming closer. "Smell it," she offers. I would not go so far as to say that her compost smells romantic, but it does not smell like rotting garbage. Compared to some of the things I've been smelling these days, it's a pot of posies.

Susanne Wiigh-Masak will not be the first person to compost a human body. That honor goes to an American named Tim Evans. I heard about Evans while visiting the University of Tennessee's human decay research facility (see Chapter 3). As a graduate student, Evans had investigated human composting as an option for third-world countries where the majority of the people can't afford coffins or cremation. In Haiti and parts of rural China, Evans told me, unclaimed bodies and bodies from poor families are often dumped in open pits. In China, the corpses are then burned using high-sulfur coal.

In 1998, Evans procured the body of a ne'er-do-well whose family had donated him to the university. "He never knew he was going to end up as compost guy," recalled Evans, when I telephoned him. This was probably just as well. To supply the requisite bacteria to break down the tissue, Evans composted the body with manure and soiled wood shavings from stables. The dignity issue rears its delicate head. (Wiigh-Masak would not be using manure; she plans to mix a "little dose" of freeze-dried bacteria in with each box of remains.)

And because the man was buried whole, Evans had to go out with a shovel and rake to aerate him three or four times. This is why Wiigh-Masak plans to break bodies up, with either vibration or ultrasound. The tiny pieces are easily saturated with oxygen and so quickly composted and assimilated that they can be used immediately for a planting. It was also,

OUT OF THE FIRE, INTO THE COMPOST BIN 265

in part, a matter of dignity and aesthetics. "The body has to be unrecognizable while it composts," says Wiigh-Masak. "It has to be in small pieces. Can you imagine the family sitting around the dinner table and someone says, 'Okay, Sven, it's your turn to go out and turn Mother'?"

Indeed, Evans had something of a rough go of it, though in his case it was more the setting than the deed. "It was hard being out there," he told me. "I used to think, 'What am I doing here?' I'd just put on my blinders and go to my pile."

It took a month and a half for compost guy to complete his return to the soil. Evans was pleased with the result, which he described as "really dark, rich stuff, with good moisture-holding capacity." He offered to send me a sample, which might or might not have been illegal. (You need a permit to ship an unembalmed cadaver across state lines, but there is nothing on the books regarding the shipping of a composted cadaver. We decided to leave it be.) Evans was pleased to note that a healthy crop of weeds had begun growing out of the top of the compost bin toward the end of the process. He had been concerned about certain fatty acids in the body, which might, if not thoroughly broken down, prove toxic to plant roots.

In the end, the government of Haiti respectfully declined Evans's proposal. The Chinese government—in what was either a remarkable show of environmental concern or a desire to save money, manure being cheaper than coal—did express interest in human composting as an alternative to open-pit coal burnings. Evans and his adviser, Arpad Vass, prepared a white paper on the practical advantages of human compost-ing (". . . material can then be safely used in land applications as a soil amendment or fertilizer") but received no further word. Evans has plans to work with veterinarians in southern California to make composting available to pet owners. Like Wiigh-Masak, he envisions families planting a tree or shrub, which would take up the deceased's molecules and become a living memorial. "This is as close," he said to me, "as science is going to get to reincarnation."

I asked Evans if he plans to try to crack the mortuary market. There are two questions there, he answered. If I was asking whether he wanted to make composting available to people, the answer was yes. But he didn't feel sure he wanted to make the process available through funeral homes. "One of the things that got me interested in this is a disdain for current practices of the funeral industry," he said. "You shouldn't have to pay exorbitant amounts of money to die." Ultimately, he'd like to offer it through a company of his own.

I then asked how he imagined he'd get the word out, get the ball rolling. He said he had tried to get a celebrity interested in the cause. The hope was that someone like Paul Newman or Warren Beatty might do for composting what Timothy Leary did for space burials. As Evans was living in Lawrence, Kansas, at the time, he called fellow Kansan William S. Burroughs, who struck him as suitably eccentric and moribund to consider it. The calls were not returned. He eventually did try to contact Paul Newman. "His daughter runs a horse stable doing rehabilitation for handicapped kids. I thought we could use the manure," Evans said. "They were probably thinking, 'What a freak.'" Evans isn't a freak. He's just a freethinker, on a topic most people would rather not think about.

Evan's adviser, Arpad Vass, summed it up best. "Composting is a wonderful possibility. I just don't think the mentality of this country is there yet."

The mentality of Sweden is a good deal closer. The thought of "living on" as a willow tree or a rhododendron bush might easily appeal to a nation of gardeners and recyclers. I don't know what percentage of Swedes have gardens, but plants seem very important to them. Business lobbies in Sweden hold tiny forests of potted trees. (In a roadside restaurant in Jönköping, I saw a ficus plant *inside* a revolving door.) The Swedes are a practical people, a people who appreciate simplicity and abhor frou-frou. The stationery of the Swedish king is simply embossed with his seal; at a distance it appears to be a plain sheet of cream-colored paper. Hotel rooms are furnished with what a reasonable traveler might need and nothing more.* There is one pad of paper, not three, and the end of the toilet paper is not triangulated. To be freeze-dried and reduced to a hygienic bag of compost and incorporated into a plant, I suppose, might appeal to the Swedish ethos.

That is not the only thing that has made Sweden the right place at the right time for the human compost movement. As it happens, the crematoria in Sweden have been hit with environmental regulations regarding volatilized mercury from fillings, and many need to make costly upgrades to their equipment within the next two years. Purchasing Wiigh-Masak's machinery would, she says, cost many of them less than would complying with government regulations. And burial hasn't been popular here for decades. Wiigh-Masak explained that part of the Swedes' distaste for interment can be traced to the fact that in Sweden you must share your grave. After twenty-five years, a grave is reopened, and "the men in gas masks," as Wiigh-Masak puts it, haul you up, dig the grave deeper, and bury someone else on top of you.

This is not to say that Promessa faces no resistance. Wiigh-Masak must convince the people whose jobs will be affected should composting become a reality: the funeral directors, the

---

* And sometimes less. My business-grade room at Gothenburg's Landvetter Airport Hotel ("For Flying People") had no clock, the assumption being, I suppose, that a businessman can simply consult his watch. The TV remote had no mute button. I pictured Swedish remote designers arguing quietly in their cleanly appointed conference room. "But Ingmar, why do you need a special button when you can just put down the volume?"

coffin makers, the embalmers. People whose apple carts stand to be upset. Yesterday she gave a talk at a conference of parish administrators in Jönköping. These are the people who would care for the person-plants in the churchyard memorial park. While she spoke, I scanned the audience for smirks and rolling eyes, but saw none. Most of the comments seemed positive, though it was hard to tell, as the comments were in Swedish and my interpreter had never actually interpreted before. He consulted frequently with a piece of graph paper, on which he had written out a list of mortuary and composting vocabulary in Swedish and English (*formultning*—"moldering, decay"). At one point, a balding man in a dark gray suit raised his hand to say that he thought composting took away the specialness of being human. "In this process, we are equal to some animal that dies in the woods," he said. Wiigh-Masak explained that she was only concerned with the body, that the soul or spirit would be addressed, as it has always been, in a memorial service or ritual of the family's choosing. He didn't seem to hear this. "Do you look around this room," he said, "and see nothing more than a hundred bags of fertilizer?" My interpreter whispered that the man was a funeral director. Apparently three or four of them had crashed the conference.

When Wiigh-Masak finished and the crowd moved to the back of the hall for coffee and pastries, I joined the man in the gray suit and his fellow undertakers. Across from me sat a man with white hair, named Curt. He wore a suit too, but his was checkered and he had an air of jollity that made it hard for me to picture him running a funeral home. He said he thought that the ecological funeral would one day, perhaps in ten years, become a reality. "It used to be that the priest told the people how to do it," he said, referring to memorial rites and rituals and the disposition of the body. "Today the people tell the priest." (According to Prothero, this was also the case with cremation. Part of the appeal of scattering ashes was that it took the last rites out of the hands of the undertakers and handed them over to the family and friends, freeing them to do something more personally meaningful than what the undertaker might have had in mind.)

Curt added that young people in Sweden had recently begun moving away from cremation because of the pollution it creates. "Now the young can go to Grandma and say, 'I have a new way for you—a cold bath!'" Then he laughed and clapped his hands. I decided that this was the sort of man I wanted running my funeral.

Wiigh-Masak joined us. "You are a very good salesman," the man in the gray suit told her. He works for Fonus, Scandinavia's largest mortuary corporation. The man let Wiigh-Masak take in the compliment before stepping on it: "But you haven't convinced me."

Wiigh-Masak didn't flinch. "I expected to get some resistance," she told him. "That's why I'm so surprised and pleased to see that almost everyone in the audience looks happy while I talk."

"Believe me, they're not," said the man pleasantly. If I didn't have an interpreter, I'd think they were discussing the pastries. "I hear what they say."

On the drive back to Lyrön, the man in the gray suit became known as The Slime.

"I hope we don't see him tomorrow," Wiigh-Masak said to me. At three o'clock the following afternoon, in Stockholm, she was scheduled to give a presentation before the top regional managers of Fonus. That she was speaking there was a matter of some pride. Two years ago, they hadn't returned her phone calls. This time it was they who called her.

Susanne Wiigh-Masak does not own a business suit. She delivers her presentations in what American dress code arbiters would term "smart-casual" trousers and a sweater, with her waist-length, wheat-colored hair braided and pinned up in back. She wears no makeup for these talks, though her face tends to flush mildly, bestowing a youthful blush.

In the past, the organic look has worked in Wiigh-Masak's favor. When she met with Church of Sweden clergy back in 1999, they were comforted by Wiigh-Masak's noncommercial mien. "They said to me, 'You are really not a seller,'" she tells me as she dresses for the trip to Fonus's Stockholm headquarters. She really isn't. While as 51 percent owner of Promessa's shares Wiigh-Masak stands to earn a substantial sum should the process take off, wealth is clearly not her motive. Wiigh-Masak has been a hard-core ecologist since the age of seventeen. This is a woman who takes trains instead of driving, to make herself less of a burden on the environment, and who disapproves of holiday-makers flying to Thailand when a beach in Spain would suffice, on the grounds that jet fuel is needlessly burned. She readily admits that Promessa has little to do with death and everything to do with the environment, that it is essentially a vehicle for spreading the gospel of ecology. The dead bodies attract the media and public attention in a way that the environmental message alone could not. She is a rarity among social advocates: the environmentalist who is not preaching to the converted. Today is a good example: Ten mortuary company executives are about to sit through an hour-long talk about the importance of giving back to the earth through organic composting. How often does *that* happen?

The Fonus headquarters takes up the better part of the third floor of a nondescript Stockholm office building. The interior designers have gone out of their way to infuse color and nature into the surroundings. An arrangement of café tables is surrounded by a sort of indoor hedge of potted trees, in the midst of which stands an immaculate tropical fish tank the size of a plate-glass window. Death is nowhere in evidence. A bowl of complimentary lint brushes bearing the Fonus logo calls out to me from the receptionist's desk.

Wiigh-Masak and I are introduced to Ulf Helsing, a vice director of the corporation. The name hits my ears as Elf Helsing, causing great internal merriment. Helsing is dressed like all the other elves in the lobby, in the same gray suit, with the same royal-blue dress shirt and the same subdued tie and silver Fonus lapel pin. I ask Helsing why Fonus instigated the meeting. As Wiigh-Masak envisions it, it is Sweden's crematoria, until recently operated by the church, that would be doing the freeze-drying. The funeral homes would simply make the option known to their clients—or not, depending on what they decide. "We have been following this in the paper, but we kept a low profile," came his enigmatic reply. "It is time we heard more." Possibly contributing to the decision was the fact that 62 percent of three hundred visitors to the Fonus Web site answered, in a survey, that they would be interested in an ecological funeral.

"You know," Helsing adds as he stirs his coffee, "that freeze-drying corpses is not a new idea. Someone in your country came up with this, about ten years ago." He is talking about a retired science teacher from Eugene, Oregon, named Phillip Backman. Wiigh-Masak told me about him. Backman, like Tim Evans and the cremationists of yore, was inspired by a loathing of funerary pomp. He spent several years at Arlington National Cemetery arranging military funerals that, much of the time, no one showed up for. This, combined with a background in chemistry, got him interested in the possibilities of freeze-drying as another alternative to burial. He knew that liquid nitrogen, a waste product of certain industrial processes, is cheaper than natural gas. (Wiigh-Masak estimates the liquid nitrogen cost per body at $30; the gas for a cremation costs about $100.) To break down the frozen bodies—for freeze-drying a whole human body would take over a year—into tiny, quickly freeze-dryable pieces, he proposed running them through a machine. "It's something on the order of what they do with chipped beef," he told me when we spoke. ("It was a *hammer mill*," Wiigh-Masak later told me.) Backman managed to secure a patent for the process, but the concept was coolly received at local mortuaries. "No one wanted to talk about it, so I just let it go."

The meeting begins on time. Ten regional directors for the company, along with their laptops and their polite gazes, gather in the conference room. Wiigh-Masak begins by talking about the difference between organic and inorganic remains, how cremains contain little nutritive value. "When we are burning remains, we don't give it back to the earth. We are built up from nature, and we have to give it back." The audience seems respectfully quiet and attentive, except for my interpreter and me, whispering in the back row like poorly brought up schoolgirls. I notice Helsing writing. At first he appears to be taking notes, but then he folds the sheet in two, and, when Wiigh-Masak's back is turned, slides it across the table, where it is passed along to its recipient, who slips it under his notebook until Wiigh-Masak turns away again.

They let Wiigh-Masak talk for twenty minutes before they begin asking questions. Helsing leads the pack. "I have an ethical question," he says. "An elk dying in the woods and returning to the earth is just lying on the ground. Here you are doing something to break it up." Wiigh-Masak replies that in fact, an elk that dies in the woods is likely to be torn up and eaten by scavengers. And while it is true that the dung of whoever eats the elk would act as a sort of elk compost and, in effect, achieve the desired goal, it was not something she could envision families being comfortable with.

Helsing pinkens slightly. This was not where he intended things to go, conversationally. He persists: "But can you see the ethical problem of breaking it up this way?" Wiigh-Masak has heard this line of argument before. A technician at a Danish ultrasound company, whom she contacted early on in the project, declined to work with her for this reason. He felt that representing ultrasound as a nonviolent way of breaking up tissue was dishonest. Wiigh-Masak was undeterred. "Listen," she said to the morticians. "We all know that taking a body down to powder requires some kind of energy. But ultrasound, at least, has a positive image. You cannot *see* the violence. I would like it to be possible for the family to watch it happening, behind a glass wall. I want something where I can show a child, and the child won't start crying." Glances are exchanged. A man clicks his pen.

Wiigh-Masak makes a small detour into defensive mode. "I think that if you put a camera inside a coffin we wouldn't be very impressed with ourselves. It is a terrible result."

Someone asks why the freeze-drying step is needed. Wiigh-Masak answers that if you don't remove the water, the little pieces will start to decompose and smell before you can get them into the ground. But you mustn't get rid of the water, the man counters, because this is 70 percent of this person. Wiigh-Masak tries to explain that the water inside each one of us changes day by day. It's borrowed. It comes in, it goes out, the molecules from your water mix with someone else's. She points to the man's coffee cup. "The coffee you are drinking has been your neighbor's urine." You have to admire a woman who can toss the word "urine" into a corporate presentation.

The man who has been clicking his pen is the first to raise the subject that is surely on everyone's mind: coffins, and the disappearing profit therefrom that an ecological funeral movement will mean. Wiigh-Masak envisions the freeze-dried, powdered remains being placed in a miniature, biodegradable cornstarch coffin. "That's a problem," acknowledges Wiigh-Masak. "Everyone will be angry at me." She smiles. "I guess there will have to be a new thinking." (As with cremation, a standard coffin could be rented for a memorial service.)

Cremationists faced the same objections. For years, according to Stephen Prothero, undertakers were advised to tell their clients that scattering was against the law, when in fact, with few exceptions, it wasn't. Families were pushed to buy memorial urns and niches in columbaria and even standard cemetery plots in which to bury the urns. But the families persisted in their push for a simple, meaningful ceremony of their own making, and scattering caught on. As did the use of rental caskets for pre-cremation services and the manufacturing of inexpensive cardboard "cremation containers" for the actual burning. "The only reason there are rental caskets," Kevin McCabe once told me, "is that the public demanded it." The tremendous attention that Promessa has received since its founding has forced the funeral industry to deal with the possibility that very soon people may be coming to them requesting to be composted. (In a Swedish newspaper poll taken last year, 40 percent of respondents said they'd like to be freeze-dried and used to grow a plant.) Mortuaries in Sweden may not be actively recommending the ecological funeral any time soon, but they may stop short of trying to derail it. As a friendly young Fonus regional director named Peter Göransson said to me earlier, "It's pretty hard to stop something once it's rolling."

The last question comes from a man seated next to Ulf Helsing. He asks Wiigh-Masak whether she plans to first market the technique for dead animals. She is adamant about not letting this happen. If Promessa becomes known as a company that disposes of dead cows or pets, she tells the man, it will lose the dignity necessary for a human application. It is difficult, as it is, to attach the requisite dignity to human composting. At least in the United States. Not long ago, I called the U.S. Conference of Catholic Bishops, the official U.S. mouthpiece of the Catholic Church, to ask its opinion on freeze-drying and composting as an alternative to burial. I was put through to a Monsignor John Strynkowski in the Doctrine office. While the monsignor allowed that composting and nourishing the earth was little different from a Trappist monk's plain shroud burial or a church-sanctioned burial at sea, following which the body will, as he put it, provide nourishment for fish, the idea of composting struck him as disrespectful. I asked him why. "Well, when I was a kid," he answered, "we had a hole where we put peelings from apples and such, and used it for fertilizer. That's just my association."

While I had him on the phone, I asked Monsignor Strynkowski about tissue digestion. He replied with minimal hesitation that the church would be opposed to "the idea of human remains going into the drain." He explained that the Catholic Church feels that the human body should always be given a dignified burial, whether it's the body itself or the ashes. (Scattering remains a sin.) When I explained that the company planned to add an optional dehydrator to the system that could reduce the liquefied remains to a powder that could then be buried, just as cremains can be, the line went quiet. Finally he said, "I guess that would be okay." You got the feeling Monsignor Strynkowski was looking forward to the end of the phone call.

The line between solid waste disposal and funerary rituals must be well maintained. Interestingly, this is one of the reasons the Environmental Protection Agency doesn't regulate U.S. crematoria. For if it did regulate them, the rules would be promulgated under Section 129 of the Clean Air Act, which covers "Solid Waste Incinerators." And that would mean, explained Fred Porter, of the EPA Emission Standards Division in Washington, "that what we're incinerating at crematoria is 'solid waste.'" The EPA does not wish to stand accused of calling America's dead loved ones "solid waste."

Wiigh-Masak may succeed in taking composting mainstream because she realizes the importance of keeping respectful disposition distinct from waste disposal, of addressing the
family's need for a dignified end. To a certain extent, of course,
dignity is in the packaging. When you get right down to it,
there is no dignified way to go, be it decomposition, incineration, dissection, tissue digestion, or composting. They're all,
bottom line, a little disagreeable. It takes the careful application
of a well-considered euphemism—burial, cremation, anatomical gift-giving, water reduction, ecological funeral—to bring it
to the point of acceptance. I used to think the traditional navy
burial at sea sounded nice; I pictured the sun on the ocean, the
infinite expanse of blue, the nowhereness of it. Then one day
I had a conversation with Phillip Backman, during which he
mentioned that one of the cleanest, quickest, and most ecologically pure things to do with a body would be to put it in a
big tidepool full of Dungeness crabs, which apparently enjoy
eating people as much as people enjoy eating crabs. "It'll do
the thing in a couple of days," he said. "It's all recycled, and it's
all clean and taken care of." My affinity for burial at sea—not
to mention crabmeat—was suddenly, dramatically diminished.

Wiigh-Masak finishes speaking, and the group applauds. If
they think of her as the enemy, they do a good job of conceal-
ing it. On the way out, a photographer asks us to pose with
Helsing and a couple of the other executives for the company
Web page. We stand with one foot and shoulder forward,
arranged in facing columns, like doo-wop backup singers in
unusually drab costumes. While I avail myself of a Fonus lint
brush, I hear Helsing say that the company plans to add a link
to Promessa on its Web site. A wary friendship has been forged.

On the road between Jönköping and Wiigh-Masak's home
on Lyrön is a graveyard on a hill. If you drive all the way

through to the back of this graveyard, you come to a small field where the church will one day dig more graves. Halfway up the unmown terrain, a small rhododendron bush stands among the weeds. This is the Promessa test grave. Last December, Wiigh-Masak concocted the approximate equivalent of a 150-pound human cadaver, using freeze-dried cow blood and freeze-dried, pulverized bones and meat. She placed the powder in a cornstarch box, and the box in a shallow (thirty-five centimeters down, so the compost could still get oxygen) grave. In June, she will return to dig it up and make sure the container has disintegrated and the contents have begun their metaphysical journey.

Wiigh-Masak and I stand in silence beside the grave of the unknown livestock, as though paying our respects. It's dark now and hard to see the plant, though it appears to be doing well. I tell Wiigh-Masak that I think it's great, this quest for an ecologically sound, meaningful memorial. I tell her I'm rooting for her, then quickly rephrase the sentiment, omitting gardening-related verbs.

And I am. I hope Wiigh-Masak succeeds, and I hope WR2 succeeds. I'm all for choices, in death as in life. Wiigh-Masak is encouraged by my support, as she has been by the support of the Church of Sweden and her corporate backers and the people who have responded positively in the polls. "It was and is," she confides as the wind shimmies the leaves on the cow's memorial shrub, "very important to feel I'm not crazy."

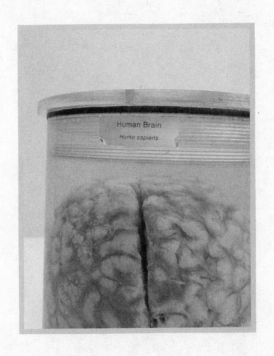

12
REMAINS OF THE AUTHOR

───────

Will she or won't she?

It has long been a tradition among anatomy professors to donate their bodies to medical science. Hugh Patterson, the UCSF professor whose lab I visited, looks at it this way: "I've enjoyed teaching anatomy, and look, I get to do it after I die." He told me he felt like he was cheating death. According to Patterson, the venerable anatomy teachers of Renaissance Padua and Bologna, as death sidled near, would choose their best student and ask him to prepare their skull as an anatomical exhibit. (Should you one day visit Padua, you can see some of these skulls, at the university medical school.)

I don't teach anatomy, but I understand the impulse. Some months back, I gave thought to becoming a skeleton in a medical school classroom. Years ago I read a Ray Bradbury story about a man who becomes obsessed with his skeleton. He has come to think of it as a sentient, sinister entity that lives inside him, biding its time until he dies and the bones slowly prevail. I began thinking about my skeleton, this solid, beautiful thing inside me that I would never see. I didn't see it becoming my usurper, but more my stand-in, my means to earthly immortality. *I've enjoyed hanging around in rooms doing nothing much, and look, I get to do it after I die.* Plus, on the off chance that an afterlife exists, and that it includes the option of home planet visitations, I'd be able to pop by the med school and finally see what my bones looked like. I liked the idea that when I was gone, my skeleton would live on in some sunny, boisterous anatomy classroom. I wanted to be a mystery in some future medical student's head: Who was this woman? What did she do? How did she come to be here?

Of course, the mystery could as easily be engendered by a more routine donation of my remains. Upward of 80 percent of the bodies left to science are used for anatomy lab dissections. Most assuredly, a lab cadaver occupies the thoughts and dreams of its dissectors. The problem, for me, is that while a skeleton is ageless and aesthetically pleasing, an eighty-year-old corpse is withered and dead. The thought of young people gazing in horror and repulsion at my sagging flesh and atrophied limbs does not hold strong appeal. I'm forty-three, and already they're doing it. A skeleton seemed the less humiliating course.

I actually went so far as to contact a facility at the University of New Mexico's Maxwell Museum of Anthropology that accepts bodies specifically to harvest the bones. I told the woman who runs it about my book and said that I wanted to come see how skeletons are made. In the Bradbury story, the protagonist ends up having his bones pulled out through his mouth, by an alien disguised as a beautiful woman. Though he was reduced to a jellyfish heap on his living-room floor, his body remained intact. No blood was spilled.

This was, of course, not the case at the Maxwell lab. I was told I would have the choice of observing one of two steps: a "cut-down" or a "pour-off." The cut-down was more or less what it sounded like. They got the bones out the only way—barring retractable and highly specialized alien mouthparts—one can: by cutting away the flesh and muscle that surrounds them. Residual meat and sinew is dissolved by boiling the bones in a solution for a few weeks, periodically pouring off the broth and replacing the solution. I pictured the young men of Padua tending to their beloved professors' heads as they simmered and bobbed. I pictured the actors in a Shakespearean theater troupe I read about last year, confronted by a dead cast member's last request that his skull be used as Yorick. People really need to think these requests through.

About a month later, I got another e-mail from the university. They were writing to tell me they had switched to an insect-based process, wherein fly larvae and carnivorous beetles perform their own scaled-down, drawn-out version of the cut-down.

I did not sign on to become a skeleton. For one thing, I don't live in New Mexico and they won't come pick you up. Also, it turns out that the university doesn't make skeletons, only bones. The bones are left unarticulated and added to the university's osteological collection.*

No one in this country, I learned, is making skeletons for medical schools. The vast majority of the world's medical school skeletons have, over the years, been imported from Calcutta. No longer. According to a June 15, 1986, *Chicago Tribune* story, India banned the export of bones in 1985, after reports surfaced of children being kidnapped and murdered for their bones and skulls. According to one story, which I desperately hope is exaggerated, fifteen hundred children per month were being killed in the state of Bihar, their bones then sent to Calcutta for processing and export. Since the ban, the supply of human bones has dwindled to almost nothing. Some come out of Asia, where, it is rumored, they are dug up from Chinese cemeteries and stolen from Cambodia's killing fields. They are old, mossy, and generally of poor quality, and for the

---

* If you live nearby, by all means donate. The Maxwell Museum holds the world's only collection of contemporary—within the last fifteen years—human bones, used to study everything from forensics to the skeletal manifestations of diseases. P.S.: Your family can go in and visit your bones, which the staff will lay out for you, though probably not in the shape of an all-together skeleton.

most part, detailed plastic skeletons have taken their place. So much for my future as a skeleton.

For similarly dumb and narcissistic reasons, I also once considered spending eternity at the Harvard Brain Bank. I wrote about it in my Salon.com column, which was disappointing for the Brain Bank's director, who assumed I would be writing a serious article about the facility's serious and very worthwhile research pursuits. Here is an abridged version of the column:

There are many good reasons to become a brain donor. One of the best is to advance the study of mental dysfunction. Researchers cannot study animal brains to learn about mental illness because animals don't get mentally ill. While some animals—cats, for example, and dogs small enough to fit into bicycle baskets—seem to incorporate mental illness as a natural personality feature, animals are not known to have diagnosable brain disorders like Alzheimer's and schizophrenia. So researchers need to study brains of mentally ill humans and, as controls, brains of normal humans like you and me (okay, you).

My reasons for becoming a donor aren't very good at all. My reasons boil down to a Harvard Brain Bank donor wallet card, which enables me to say "I'm going to Harvard" and not be lying. You do not need brains to go to the Harvard Brain Bank, only a brain.

One fine fall day, I decided to visit my final resting place. The Brain Bank is part of Harvard's McLean Hospital, which sits on a rolling estate of handsome brick buildings just outside Boston. I was directed to the third floor of the Mailman Research Building. The woman pronounced it "Melmon," so as to avoid having to answer stupid questions about what kind of research is being done on mailmen.

If you are considering becoming a brain donor, the best thing for you to do is stay away from the Brain Bank. Within ten minutes of arriving, I was watching a twenty-four-year-old technician slice a sixty-seven-year-old brain. The brain had been flash-frozen and did not slice cleanly. It sliced as does a Butterfinger, with little shards crumbling off. The shards quickly thawed and looked less Butterfingerlike. The technician wiped them up with a paper towel. "There goes third grade." He has gotten in trouble for saying things like this. I read a newspaper story in which the reporter asked him if he planned to donate his brain and he replied, "No way! I'm going out with whatever I came in with!" Now when you ask him, he says quietly, "I'm only twenty-four, I really don't know."

A Brain Bank spokesman showed me around. Down the hall from the dissection room was the computer room. The spokesman referred to this as "the brains of the operation," which in any other operation would have been fine, but in this case was a tad confusing. At the end of the hall were the real brains. It wasn't quite what I imagined. I had pictured whole intact brains floating in glass jars. But the brains are cut in half, one side being sliced and frozen, the other side sliced and stored in formaldehyde inside Rubbermaid and Freezette food savers. Somehow, I'd expected more of Harvard. If not glass, at least Tupperware. I wondered what the dorms look like these days.

. . . The spokesman assured me that no one would even be able to tell that my brain was missing. He assured me in a way that assured me and at the same time didn't bring me a lot closer to being a committed brain donor. "First," he began, "they cut the skin like this and pull it up over the face." Here he made a motion as though taking off a Halloween mask. "They use a saw to cut the top of the skull off, the brain is removed, and the skull is put back and screwed in place. Put the skin flap back, and comb the hair back over." He used the peppy how-to language of an infomer-

cial host, making brain harvesting sound like something that
takes just minutes and wipes clean with a damp cloth. . . .

Yet again, I backed off from my plan. Not so much because
of the harvesting process—as you may have gleaned, I'm not
a squeamish individual—but because of my mistaken expecta-
tions. I wanted to be a brain in a jar, at Harvard. I wanted to
look atmospheric and fascinating on a shelf. I didn't want to
spend the hereafter as cut-up pieces in a storeroom refrigerator.

There is but one way to be an organ on a shelf, and that is
to be plastinated. Plastination is the process of taking organic
tissue—a rosebud, say, or a human head—and replacing the
water in it with a liquid silicone polymer, turning the organism into a permanently preserved version of itself. Plastination was developed by German anatomist Gunther von
Hagens. Like most plastinators, von Hagens makes educational models for anatomy programs. He is best known, however, for his controversial plastinated whole-body art exhibit,
"Körperwelten"—or, in England, "Bodyworlds"—which has
toured Europe for the past five years, raising eyebrows and
tidy sums of cash (attendance to date is over eight million).
The skinless bodies are posed as living people in action: swim-
ming, riding (plastinated horse included), playing chess. One
figure's skin flies out behind it like a cape. Von Hagens cites
as inspiration the works of Renaissance anatomists such as
Andreas Vesalius, whose *De Humani Corporis Fabrica* featured
bodies drawn in active human poses, rather than lying flat or
standing arms to the side, à la the typical medical illustration.
A skeleton waves hello; a "muscle man" gazes at the view from
a hilltop of the town below. "Körperwelten" raises the ire of
church fathers and conservatives wherever it opens, mainly on
the grounds of violated dignity. Von Hagens counters that the
bodies in the show were donated by their owners specifically
for this purpose. (He leaves a stack of donor forms at the exit

of the exhibit. According to a 2001 London *Observer* article, the donor list is up to 3,700.)

Most of von Hagens's bodies are plastinated in China, in an operation called Plastination City. He is said to employ two hundred Chinese in what sounds to me like a sort of cadaver sweat shop. This is not all that surprising, as his technique is extremely labor-intensive and time-consuming—it takes over a year to plastinate one individual. (The U.S. version of the technique, modified by Dow Corning after von Hagens's patent expired, takes one tenth the time.) I contacted von Hagens's office in Germany to see if I could visit Plastination City and see what kind of shenanigans are in store for a donor body, but von Hagens was on the road and did not return my e-mails in time.

Instead of China, I traveled to the University of Michigan Medical School, where anatomy professor Roy Glover and plastination chemicals manufacturer Dan Corcoran, who worked with Dow Corning to update the technique, have been plastinating whole dead bodies for a museum project of their own, called "Exhibit Human: The Wonders Within"— slated to open in San Francisco in mid-2003. Theirs is strictly educational: twelve plastinated (Corcoran prefers the term "polymer-preserved") bodies, each displaying a different system—nervous, digestive, reproductive, etc. (At press date, no U.S. museum had signed up to exhibit "Körperwelten.")

Glover offered to show me how plastination works. We met in his office. Glover has a long face that made me think of Leo G. Carroll. (I had recently seen *Tarantula*, wherein Carroll plays a scientist who figures out how to make huge, scary versions of harmless animals, e.g. "Guinea pigs the size of police dogs!") You could tell Glover was a nice guy because a To Do list on a white board on his office wall said: "Maria Lopez, brain for daughter—science fair." I decided that this was what I wanted to do with my remains. Travel around to classrooms and science fairs, astounding children and inspiring careers in science. Glover took me across the hall, to a storeroom with a wall of shelves crowded with plastinated human pieces and parts. There was a brain sliced like a loaf of bread and a head split in two so that you could see the labyrinths of the sinuses and the deep, secret source of the tongue. You could pick the organs up and marvel at them, for they were completely dry and had no smell. Yet still, they were clearly real and not plastic. For the many disciplines (dentistry, nursing, speech pathology) that study anatomy but have no time for dissection, models like these are a godsend.

Glover took me down the hall to the plastination lab, which was chilly and cluttered with heavy, strange-looking tanks. He began explaining the process. "First the body is washed." This is done much as it was when the body was alive: in a tub. "This is a body," said Glover, quite unnecessarily, regarding a figure on its back in the tub.

The man had been in his sixties. He had a mustache and a tattoo, both of which would survive the plastination process. The head was submerged, giving the corpse a disconcerting murder-victim sort of look. Also, the front chest wall had been separated from the rest of the torso and lay off to the side of the body. It looked like a Roman gladiator's chest plate, or maybe I just found it helpful to think of it that way. Glover said that he and Corcoran planned to reattach it with a hinge on one side, so that it would swing open "like a refrigerator door" to reveal the organs within. (Months later, I saw photos of the exhibit pieces. Disappointingly, someone must have nixed the refrigerator door idea.)

The second body lay in a stainless-steel tank of acetone, which filled the lab with a powerful smell of nail polish remover each time Dr. Glover lifted the lid. The acetone drives water from the body's tissue, readying it for impregnation with the silicone polymer. I tried to picture this dead man propped on a stand in a science museum. "Will he be wearing anything, or will his penis just be hanging out?" I asked tactlessly.

"He's going to have it hanging out," replied Glover. I got the feeling he'd been asked this question before. "I mean, this is a perfectly normal part of a person's anatomy. Why should we attempt to hide what's normal?"

From the acetone bath, the cadavers are transferred to the whole-body plastination chamber, a cylindrical stainless-steel tank filled with liquid polymer. A vacuum attached to the tank lowers the internal pressure, turning the acetone to a gas and drawing it from the body. "When the acetone comes out of the specimen, it creates space, and into that space is pulled the polymer," said Glover. He handed me a flashlight so I could see the view through a porthole on the top of the chamber, which happened to look down onto a perfectly normal part of a person's anatomy.

It looked peaceful in there. Like a guinea pig the size of a police dog, the concept of being plastinated is more unsettling than the reality. You just lie there, soaking and plastinating. Eventually, someone lifts you out and poses you, much as one poses a Gumby. A catalyst is then rubbed into your skin, and a two-day hardening process begins, working its way through your tissues, preserving you for all eternity in your freshly dead state. I asked Dean Mueller, a southeastern Michigan funeral director whose company, Eternal Preservation, offers mortuary plastination for about $50,000, how long he thought a plastinated specimen would last. He said at least ten thousand years, which is about as eternal as anyone in their right, or even their wrong, mind could care about. Mueller has high hopes that the process will catch on among heads of state (think what plastination could have done for Lenin) and rich eccentrics, and I imagine that it might.

I would happily donate my organs as teaching tools, but unless I move to Michigan or some other state with a plastination lab, I can't. I could ask my loved ones to ship me to Michigan, but that would be silly. Besides, you can't specify what happens to you when you donate your remains to science, only what doesn't happen. The dead people whose parts Glover and Corcoran have plastinated over the years checked a box on their University of Michigan donor form indicating that they did not object to "permanent preservation," but they didn't request it specifically.

Here's the other thing I think about. It makes little sense to try to control what happens to your remains when you are no longer around to reap the joys or benefits of that control. People who make elaborate requests concerning disposition of their bodies are probably people who have trouble with the concept of not existing. Leaving a note requesting that your family and friends travel to the Ganges or ship your body to a plastination lab in Michigan is a way of exerting influence after you're gone—of still being there, in a sense. I imagine it is a symptom of the fear, the dread, of being gone, of the refusal to accept that you no longer control, or even participate in, anything that happens on earth. I spoke about this with funeral director Kevin McCabe, who believes that decisions concerning the disposition of a body should be made by the survivors, not the dead. "It's none of their business what happens to them when they die," he said to me. While I wouldn't go that far, I do understand what he was getting at: that the survivors shouldn't have to do something they're uncomfortable with or ethically opposed to. Mourning and moving on are hard enough. Why add to the burden? If someone wants to arrange a balloon launch of the deceased's ashes into inner space, that's fine. But if it is burdensome or troubling for any reason, then perhaps they shouldn't have to. McCabe's policy is to honor the wishes of the family over the wishes of the dead. Willed body program coordinators feel similarly. "I've had kids object to their dad's wishes [to donate]," says Ronn Wade, director of the Anatomical Services Division of the University of Maryland School of Medicine. "I tell them, 'Do what's best for you. You're the one who has to live with it.'"

I saw this happen between my father and mother. My father, who rejected organized religion early in his life, asked my mother to have him cremated in a plain pine box and to hold no memorial service. My mother, against her Catholic inclinations, honored his wishes. She later regretted it. People she barely knew confronted her about their disappointment over there having been no memorial service. (My father had been a beloved character around town.) My mother felt shamed and slandered. The urn was a further source of discomfort, partly because the Catholic Church insists on burial of remains, even cremated ones, and partly because she didn't like having it around the house. Pop sat in a closet for a year or two until one day, with no word to my brother or me, she brought him down to the Rand Funeral Home, pushed aside her guilt, and had the urn buried in a cemetery plot beside the one she'd reserved for herself. Initially, I had sided with my father and was indignant over her disrespect of his stated request. When I realized how distressing his last wishes had been for her, I changed my mind.

If I donated my body to science, my husband, Ed, would have to picture me on a lab table and, worse, picture all the things that might be done to me there. Many people would be fine with this. But Ed is squeamish about bodies, living or dead. This is a man who refuses to wear contacts because he'd have to touch his eyes. I have to limit my visits to the Surgery Channel for evenings when he's out of town. When I told him I was thinking about joining the Harvard Brain Bank a couple years back, he started shaking his head: "Ix-nay on the ainbank-bray."

Whatever Ed wants to do with me is what will be done with me. (The exception being organ donation. If I wind up brain-dead with usable parts, someone's going to use them, squeamishness be damned.) If Ed goes first, only then do I fill out the willed body form.

And if do, I will include a biographical note in my file for the students who dissect me (you can do this), so they can look down at my dilapidated hull and say, "Hey, check this. I got that woman who wrote a book about cadavers." And if there's any way I can arrange it, I'll make the thing wink.

Epilogue

2021

Some things haven't changed. People still die, and some of them—more now than when I wrote the book, happily—choose to donate their bodies to science. For the majority of these good people, their remains will wind up on a dissecting table, helping students learn anatomy. While virtual anatomy software has become more sophisticated in the years since *Stiff* came out, it hasn't replaced hands-on dissection. In the words of one marginally impressed anatomy professor, "There are so many really expensive things out there that are not worthwhile at all."

Still, dissection labs take up a tremendous amount of time, of which students have less and less as medical school curricula continue to expand. Because of these time constraints, gross anatomy courses today tend to offer a combination of dissection and virtual learning. COVID-19 may have pushed the trend toward the latter. The lockdowns of 2020 sent most gross anatomy students fully online, and budget-conscious administrators may be tempted to continue the practice. It will depend on how well the students perform. In the early 2000s, a few schools experimented with eliminating dissection from anatomy courses. Board exam scores dropped, and cadavers returned to campus.

Memorial services like the one I attended at UCSF are now the norm. One school, the University of Oklahoma College of Medicine, has taken things a step further. At a luncheon on campus, new anatomy students are introduced to family members of the actual donor they will shortly begin to take apart. When a research team at the college solicited feedback from some of the families, almost all of it was positive. Many of the relatives were moved to become donors themselves. (Though probably not anyone from the family who sat down to lunch with a group of students who seemed "more interested in where to go to get the bus to get out of there than they were about Mom.")

With the growth of computerized anatomy instruction, body donation programs are having to update their consent forms. Virtual dissection and anatomy instruction software are for-profit undertakings, and not everyone may feel comfortable with images of their dead selves existing in cyberspace.

Consent forms are also adding clauses to cover genetic research. One reason: Scientists recently figured out how to recover DNA from embalmed human tissue. To look at an individual's genes, it's necessary to unwind the DNA, and embalming prevented that—until now. (Embalming chemicals keep tissue from breaking down by creating extra bonds between proteins.) It's an exciting development, as it means that researchers can search large numbers of embalmed bodies for specific genes associated with diseases—Alzheimer's, say, or lumbar disk degeneration. They can then examine the brains or spines of those bodies to determine whether the individuals with those genes were in fact more likely to have developed the disease, how much more likely, and what sort of damage or risks the gene actually appears to create.

Vehicular safety continues to be a vital postmortem career. My later books followed cadavers into safety tests of splash-landing space capsules and armored personnel carriers that need to keep soldiers safe should a bomb explode under the vehicle. Both are scenarios for which existing front-impact or side-impact crash test dummies can't provide relevant data.

Self-driving cars and trucks will present more safety unknowns. If drivers no longer need to face forward in their seats, how should manufacturers configure the safety restraints? How do you protect the occupants of a car if they're stretched out napping or sitting in a circle having lunch? Where should the airbags go? Companies initially dismissed these concerns on the grounds that their sensors and automated braking and steering systems would be so good that the cars would never crash. But it's likely to be a decade or more before all cars on the road are fully autonomous, and in the meantime, the ones with drivers will still be plowing into things, including self-driving cars. And computerized braking and steering systems, like any computers, are bound to malfunction from time to time. What happens then? "As more and more of these cars got on the road, it was like, 'Oh, guess what, one drove into a semi,'" says John Bolte, director of the Injury Biomechanics Research Center at Ohio State University. "And 'Oh, look here, one ran over a pedestrian.'" Flying taxis will also require cadaver work to calibrate the crash test dummies that will inform the vehicles' safety features and industry regulations. Safe to say, Bolte—and the dead—won't be out of work anytime soon.

Another new job opportunity for the dead: delivery-drone safety testing. Bolte built a sort of controlled slingshot system to replicate the force with which a falling drone might crash into the head of someone unlucky enough to be under it. (This has happened at least twice.) The drones are lightweight and the plastic crumples nicely, absorbing the energy of the impact such that a direct hit, he found, should be survivable. Should a drone drop a five-pound box on your head, however, all bets are off. More testing to follow.

There's good news from the world of organ transplantation. Portable perfusion machines are keeping donated organs oxygenated and viable for longer periods of time, greatly expanding the circle of possible recipients. A Hawaiian heart recently flew to North Carolina for transplantation, something that would not have been possible with the old organs-on-ice method.

Though the shortfall of needed organs is smaller than it was when I wrote this book, there still aren't enough. In 2019, more than seven thousand U.S. patients died waiting for an organ. Some European countries have enacted a presumed-consent model, wherein citizens who don't want to make their organs available to save others' lives must make the effort to opt out of the organ-sharing network. Here in the United States, we require willing donors to make the effort—to register and opt in. I asked a spokesperson at the United Network for Organ Sharing (UNOS) whether there's talk of shifting to presumed consent. The short answer is no. The slightly longer one is: This a nation where significant numbers of people will take to the streets to protest a mandate to put on a cloth mask to help prevent the spread of COVID-19. You can imagine what would happen if the families of a deceased individual who forgot to opt out later learned that their loved one's organs had been removed.

Also new: There's now a national registry for organ and tissue donation: www.registerme.org. When you finish reading this, you can sign up. Pretty please.

As with terms for people with disabilities, words and descriptors for those with the ultimate disability evolve quickly. Anatomy professionals prefer "donor" to "cadaver," and carcasses are "animal mortalities." Many universities now use "body donation program" rather than "willed body program," because you can't donate your body to science via your will. The vehicle safety people have adopted an acronym: PMHS, for postmortem human subject. In organ transplantation circles, the term "beating-heart cadaver" is no longer used. The current term is "deceased donor." It refers not only to brain-dead patients on respirators but also to terminally ill patients whose life support is removed in an operating room so that their organs can be procured ("harvested" is also passé).

Robert White's 1970s vision of attaching an entire donor body to the head of a living but terminally ill patient has not come to pass—despite the proclamations of Italian surgeon Sergio Canavero. In 2015, Canavero announced that he had a willing volunteer and was gearing up to attempt the operation within two years. The volunteer has since changed his mind, and Canavero's academic affiliation has been revoked.

I checked in with Brandy Schillace, editor in chief of the journal *Medical Humanities* and author of a new book about White and his work. What did she make of the video clips posted by Canavero's colleague Xiaoping Ren showing mice managing to drag themselves around after their spinal nerves were allegedly severed and then repaired? For one thing, she said, gross motor control is wired differently in four-legged animals. Mice are able to overcome paralysis to a degree.

Advances with prosthetic limbs suggest that it may be possible to bypass damaged spinal nerves rather than repairing or regrowing them. If the brain can learn to move a prosthetic body part, perhaps one day it could operate an entire donor body. For now, the proposition is highly sci-fi—with an emphasis on the "fi."

Moving on to body farms. As of 2021, there are seven in the United States, as well as several in Canada and Australia. They're like Starbucks! The additional sites have made it possible to fine-tune the time line of decomposition in widely varying climates. Post–*Stiff*, the research has moved on to refining the forensics of skeletal remains. New methods help determine how long bones have lain in place—either by examining decomposition chemicals in the soil below or by identifying the tooth marks of scavenging wildlife. Rats, for instance, gnaw bones while they're still relatively fresh and greasy, while squirrels arrive a year or more postmortem to chew on dry bones (seeking calcium, it's thought, for their litters). In the field of biometrics, researchers have demonstrated that identification of a dead body via iris or fingerprint patterns may be possible for longer than previously thought: up to four days postmortem in summer and fifty or more days in winter.

Mortuary composting has at last arrived in the United States. In 2020, Washington state created a new regulatory category—"reduction of human remains"—that covers cremation, natural organic reduction (composting), and alkaline hydrolysis. Colorado legalized natural organic reduction in May 2021, and California, Oregon, and Hawaii are poised to follow suit. The nation's first human composting facility, opened in 2021, has no connection to Promessa, the Swedish outfit profiled in *Stiff*. It belongs to the American company Recompose, which plans to offer franchises in other states. Promessa still exists but has yet to open its first facility, the company's Bjorn-Toni Bakken said in an e-mail. Progress, she said, has been hobbled by bankruptcies, infringements, and attempted "coups." Sadly, founder Susanne Wiigh-Masak died in 2020 (and was not composted). Their efforts continue.

Alkaline hydrolysis, the other new eco-friendly "dissolution process," has made strides as an option for livestock and pets but has been slow to take off as a mortuary option. Though AH is, as of early 2021, legal in eighteen states, fewer than 1 percent of America's dead, according to Cremation Association of North America director Barbara Kemmis, end up in a BioLiquidator (as one company calls its pressure cooker), and fewer than forty practitioners offer the service.

AH gets lumped in with cremation for regulatory purposes, but the end product—the dried and crushed "bone shadows"—

is a little different. It resembles powdered sugar and, as with most powders, if you throw it onto water, it lingers as a film on the liquid's surface—marring somewhat the poetic finality of a scattering at sea. Kemmis knows the boat owner who scatters the hydrolyzed remains of UCLA cadavers off the coast of southern California. He has to run the boat in circles for a minute to roil the surface and submerge the powder. On one occasion, he was stopped by the Coast Guard, following a report of suspicious activity called in by someone at Camp Pendleton, where the Marine Corps practices amphibious assaults along the shore. You can imagine the concern: Random guy in a boat stops in the middle of nowhere and upends an unmarked drum of white powder into the water.

As you are likely aware, Mehmet Oz left the operating room for the television studio. I met him on the set of his show during my *Gulp* book tour and found it hard to recognize the witty, skeptical heart transplanter I'd met in 2001. While we chatted, a billboard-size video screen behind us displayed an animation of a gigantic, realistic turd suddenly lurching along in a colon. Which somewhat made up for things.

Last but not least, a Pub Med search of "clitoris" and "cadaver" now brings up forty-three anatomy papers, not just one. Huzzah!

On a personal note, I am finally filling out the paperwork for a university body donation program. Along with it, I plan to submit a letter for the medical students or researchers in whose service my dead self will one day wind up:

Hello there. Pardon my appearance. I have a pretty good sense of how I must look, because I spent some time around cadavers for a book I wrote. I thought I'd share a little background about the pale form with which you'll soon be intimately acquainted. At sixty-one, as I write this, all my organs seem to be working fine. Who knows which ones will one day betray me. That's for you to discover.

I was always pretty scrawny. I once won first prize in a knobby knee contest. My right knee invariably and for no discernible reason sets off airport metal detectors. Have a look, will you? Perhaps aliens implanted something without my knowledge. I'm kidding there. When you open up my brain, you won't find any gray matter devoted to UFOs or astrology or conspiracy theories. Just facts and memories, insecurities and everyday anxieties. And the lyrics to every pop hit and television theme song from the 1970s.

My feet served me well and without complaint. (I rarely wore heels and hope no one feels they have to anymore.) They wandered around on tundra moss and lava beds, on the hard snow of an Antarctic ice field, the command deck of a ballistic missile submarine, the worn linoleum of obscure and far-flung research labs. I logged millions of miles with this body, and I hope you do the same with yours. The world is astonishing.

I'll never know the particulars of what you'll be doing with my cold hull, but I trust it will be educational. All that is good and noble in this world begins with education. Learn well and live well. And if you post pictures of me on the Internet, I will track you down in hell.

Acknowledgments

People who work with cadavers do not, as a general rule, enjoy the spotlight. Their work is misunderstood and their funding vulnerable to negative publicity. What follows is a group of people who had every reason not to return my calls, yet did. Commander Marlene DeMaio, Colonel John Baker, and Lieutenant Colonel Robert Harris, I salute your candor. Deb Marth, Albert King, John Cavanaugh, and the staff of the Wayne State impact lab, thank you for opening doors that don't often get opened. Rick Lowden, Dennis Shanahan, Arpad Vass, and Robert White, thank you for being charming and endlessly patient while I asked inane questions and used up entire afternoons of your time.

For helping make impossible things possible, I must thank the miraculous Sandy Wan, John Q. Owsley, Von Peterson, Hugh Patterson, and my pal Ron Walli. An especially warm thank-you to Susanne Wiigh-Masak and her family for putting up with me (and putting me up) for three days and nights. For sharing their time and tremendous knowledge, I thank Cindy Bir, Key Rey Chong, Dan Corcoran, Art Dalley, Nicole D'Ambrogio, Tim Evans, Roy Glover, John T. Greenwood,

Don Huelke, Paul Israel, Gordon Kaye, Tyler Kress, Duncan MacPherson, Aris Makris, Theo Martinez, Kevin McCabe, Mack McMonigle, Bruce Latimer, Mehmet Oz, Terry Spracher, Jack Springer, Dennis Tobin, Ronn Wade, Mike Walsh, Med-O Whitson, Meg Winslow, and Frederick Zugibe.

A big hug to Jeff Greenwald for the support and martinis, to Laura Fraser for her unflagging enthusiasm, and to Steph Gold, who spent three days of her summer vacation with me in Haikou, China, when almost anywhere else would have been more fun. I thank Clark for being Clark, Lisa Margonelli for making me laugh when all was darkest, and Ed for loving a woman who writes about cadavers.

Special thanks must go to David Talbot, brave and brilliant founder of Salon.com, for getting the ball rolling, and to my smart and outrageously good agent, Jay Mandel. To my editor, the gifted poet and novelist Jill Bialosky, thank you endlessly for your patience, vision, and editorial grace. Every writer should be so fortunate.

And finally, my gratitude to UM 006, H, Mr. Blank, Ben, the big guy in the sweatpants, and the owners of the forty heads. You are dead, but you're not forgotten.

Bibliography

CHAPTER 1: A HEAD IS A TERRIBLE THING TO WASTE

Burns, Jeffrey P., Frank E. Reardon, and Robert D. Truog. "Using Newly Deceased Patients to Teach Resuscitation Procedures." *New England Journal of Medicine* 331 (24):1652–55 (1994).

Hunt, Tony. *The Medieval Surgery*. Rochester: Boydell Press, 1992.

The Lancet. "Cooper v. Wakley." 1828–29 (1), 353–73.

———. "Guy's Hospital." 1828–29 (2), 537–38.

Richardson, Ruth. *Death, Dissection, and the Destitute*. London: Routledge & Kegan Paul, 1987.

Wolfe, Richard J. *Robert C. Hinckley and the Recreation of the First Operation Under Ether*. Boston: Boston Medical Library in the Francis A. Countway Library of Medicine, 1993.

CHAPTER 2: CRIMES OF ANATOMY

Bailey, James Blake. *The Diary of a Resurrectionist*. London: S. Sonnenschein, 1896.

Ball, James Moores. *The Sack-'Em-Up Men: An Account of the Rise and Fall of the Modern Resurrectionists*. London and Edinburgh: Oliver & Boyd, 1928.

Berlioz, Hector. *The Memoirs of Hector Berlioz*. Edited by David Cairns. London: Victor Gollancz, 1969.

Cole, Hubert. *Things for the Surgeon: A History of the Resurrection Men.* London: Heinemann, 1964.

Dalley, Arthur F., Robert E. Driscoll, and Harry E. Settles. "The Uniform Anatomical Gift Act: What Every Clinical Anatomist Should Know." *Clinical Anatomy* 6:247–54 (1993).

The Lancet. "Human Carcass Butchers." Editorial, 31 January 1829. 1828–29 (1), 562–63.

———. "The Late Horrible Murders in Edinburgh, to Obtain Subjects for Dissection." Abridged from *Edinburgh Evening Courant.* 1828–29 (1), 424–31.

Lassek, A. M. *Human Dissection: Its Drama and Struggle.* Springfield, Ill.: Charles C. Thomas, 1958.

O'Malley, C. D. *Andreas Vesalius of Brussels 1514–1564.* Berkeley and Los Angeles: University of California Press, 1964.

Onishi, Norimitsu. "Medical Schools Show First Signs of Healing from Taliban Abuse." *New York Times,* 15 July 2002, A10.

Ordoñez, Juan Pablo. *No Human Being Is Disposable: Social Cleansing, Human Rights, and Sexual Orientation in Colombia.* A joint report of the Colombia Human Rights Committee, the International Gay and Lesbian Human Rights Commission, and Proyecto Dignidad por los Derechos Humanos en Colombia, 1995.

Persaud, T. V. N. *Early History of Human Anatomy: From Antiquity to the Beginning of the Modern Era.* Springfield, Ill.: Charles C. Thomas, 1984.

Posner, Richard A., and Katharine B. Silbaugh. *A Guide to America's Sex Laws.* Chicago: University of Chicago Press, 1996.

Rahman, Fazlur. *Health and Medicine in the Islamic Tradition: Change and Identity.* New York: Crossroad, 1987.

Richardson, Ruth. *Death, Dissection, and the Destitute.* London: Routledge & Kegan Paul, 1987.

Schultz, Suzanne M. *Body Snatching: The Robbing of Graves for the Education of Physicians in Early Nineteenth Century America.* Jefferson, N.C.: McFarland, 1991.

Yarbro, Stan. "In Colombia, Recycling Is a Deadly Business." *Los Angeles Times,* 14 April 1992.

CHAPTER 3: LIFE AFTER DEATH

Evans, W. E. D. *The Chemistry of Death*. Springfield, Ill.: Charles C. Thomas, 1963.

Mayer, Robert G. *Embalming: History, Theory, and Practice*. Norwalk, Conn.: Appleton & Lange, 1990.

Mitford, Jessica. *The American Way of Death*. New York: Simon & Schuster, 1963.

Nhất Hanh, Thích. *The Miracle of Mindfulness*. Boston: Beacon Press, 1987.

Quigley, Christine. *The Corpse: A History*. Jefferson, N.C.: McFarland, 1996.

Strub, Clarence G., and L. G. "Darko" Frederick. *The Principles and Practice of Embalming*. 4th edition. Dallas: L. G. Frederick, 1967.

CHAPTER 4: DEAD MAN DRIVING

Brown, Angela K. "Hit-and-Run Victim Dies in Windshield, Cops Say." *Orlando Sentinel*, 3 August 2002.

Claes, H., B. Bijnenes, and L. Baert. "The Hemodynamic Influence of the Ischiocavernosus Muscles on Erectile Function." *Journal of Urology* 156:986–90 (September 1996).

Droupy, S., et al. "Penile Arteries in Humans." *Surgical Radiologic Anatomy* 19:161–67 (1997).

Edwards, Gillian M. "Case of Bulimia Nervosa Presenting with Acute, Fatal Abdominal Distension." Letter to the editor in *The Lancet*, April 6, 1985. 822–23.

King, Albert I. "Occupant Kinematics and Impact Biomechanics." In *Crashworthiness of Transportation Systems: Structural Impact and Occupant Protection*. Netherlands: Kluwer Academic Publishers, 1997.

———, et al. "Humanitarian Benefits of Cadaver Research on Injury Prevention." *Journal of Trauma* 38 (4):564–69 (1995).

Le Fort, René. *The Maxillo-Facial Works of René Le Fort*. Edited and translated by Hugh B. Tilson, Arthur S. McFee, and Harold P. Soudah. Houston: University of Texas Dental Branch.

Matikainen, Martii. "Spontaneous Rupture of the Stomach." *American Journal of Surgery* 138: 451–52.

O'Connell, Helen E., et al. "Anatomical Relationship Between Urethra and Clitoris." *Journal of Urology* 159:1892–97 (June 1998).

Patrick, Lawrence. "Forces on the Human Body in Simulated Crashes." In *Proceedings of the Ninth Stapp Car Crash Conference—October 20–21, 1965.* Minneapolis: University of Minnesota, 1966.

———. "Facial Injuries—Cause and Prevention." In *The Seventh Stapp Car Crash Conference—Proceedings.* Springfield, Ill.: Charles C. Thomas, 1963.

———, ed. *Eighth Stapp Car Crash and Field Demonstration Conference.* Detroit: Wayne State University Press, 1966.

Schultz, Willibrord W., et al. "Magnetic Resonance Imaging of Male and Female Genitals During Coitus and Female Sexual Arousal." *British Medical Journal* 319:1596–1600 (1999).

Severy, Derwyn, ed. *The Seventh Stapp Car Crash Conference—Proceedings.* Springfield, Ill.: Charles C. Thomas, 1963.

U.S. House Committee on Interstate and Foreign Commerce. *Use of Human Cadavers in Automobile Crash Testing: Hearing Before the Subcommittee on Oversight and Investigations.* 95th Cong., 2d sess. 4 August 1978.

Vinger, Paul F., Stefan M. Duma, and Jeff Crandall. "Baseball Hardness as a Risk Factor for Eye Injuries." *Archives of Ophthalmology* 117:354–58 (March 1999).

Yang, Claire, and William E. Bradley. "Peripheral Distribution of the Human Dorsal Nerve of the Penis." *Journal of Urology* 159:1912–17 (June 1998).

CHAPTER 5: BEYOND THE BLACK BOX

Clark, Carl, Carl Blechschmidt, and Fay Gorden. "Impact Protection with the 'Airstop' Restraint System. In *The Eighth Stapp Car Crash and Field Demonstration Conference—Proceedings.* Detroit: Wayne State University Press, 1966.

Mason, J. K., and W. J. Reals, eds. *Aerospace Pathology.* Chicago: College of American Pathologists Foundation, 1973.

———, and S. W. Tarlton. "Medical Investigation of the Loss of the Comet 4B Aircraft, 1967." Lancet, March 1, 1969, 431–34.

Snyder, Richard G. "Human Survivability of Extreme Impacts in Free-Fall." Civil Aeromedical Research Institute, August 1963. Reproduced by the National Technical Information Service, Springfield, Va., publication AD425412.

Synder, Richard G., and Clyde C. Snow. "Fatal Injuries Resulting from Extreme Water Impact." Civil Aeromedical Institute, September 1968. Reproduced by the National Technical Information Service, Springfield, Va., publication AD688424.

Vosswinkel, James A., et al. "Critical Analysis of Injuries Sustained in the TWA Flight 800 Midair Disaster." *Journal of Trauma* 47 (4):617–21.

Whittingham, Sir Harold, W. K. Stewart, and J. A. Armstrong. "Interpretation of Injuries in the Comet Aircraft Disasters." *Lancet*, June 4, 1955, 1135–44.

CHAPTER 6: THE CADAVER WHO JOINED THE ARMY

Bergeron, D. M., et al. "Assessment of Foot Protection Against Anti-Personnel Landmine Blast Using a Frangible Surrogate Leg." UXO Forum 2001, 9–12 April 2001.

Fackler, Martin L. "Theodor Kocher and the Scientific Foundation of Wound Ballistics." *Surgery, Gynecology & Obstetrics* 172:153–60 (1991).

Göransson, A. M., D.H. Ingvar, and F. Kutyna. "Remote Cerebral Effects on EEG in High-Energy Missile Trauma." *Journal of Trauma*, January 1988, S204.

Haller, Albrecht von. *A Dissertation on the Sensible and Irritable Parts of Animals.* Baltimore: Johns Hopkins Press, 1936.

Harris, Robert M., et al. *Final Report of the Lower Extremity Assessment Program (LEAP).* Vol. 2, USAISR Institute Report No. ATC-8199, August 2000.

Jones, D. Gareth. *Speaking for the Dead: Cadavers in Biology and Medicine.* Brookfield, England: Ashgate, 2000.

La Garde, Louis A. *Gunshot Injuries: How They Are Inflicted, Their Complications and Treatment.* New York: William Wood, 1916.

Lovelace Foundation for Medical Education and Research. *Estimate of Man's Tolerance to the Direct Effects of Air Blast.* Defense Atomic Support Agency Report, October 1968.

MacPherson, Duncan. *Bullet Penetration: Modeling the Dynamics and the Incapacitation Resulting from Wound Trauma.* El Segundo, Calif.: Ballistic Publications, 1994.

Marshall, Evan P., and Edwin J. Snow. *Handgun Stopping Power: The Definitive Study.* Boulder, Colo.: Paladin Press, 1992.

Phelan, James M. "Louis Anatole La Garde, Colonel, Medical Corps, U.S. Army." *Army Medical Bulletin* 49 (July 1939).

Surgeon General of the Army. "Report of Capt. L. A. La Garde." *Report to the Secretary of War for the Fiscal Year 1893.* Washington: Government Printing Office, 1893.

U.S. Senate. *Transactions of the First Pan-American Medical Congress.* 53rd Cong., 2d sess., Part I. 5, 6, 7, and 8 September 1893.

CHAPTER 7: HOLY CADAVER

Barbet, Pierre. *A Doctor at Calvary: The Passion of Our Lord Jesus Christ as Described by a Surgeon.* Fort Collins, Colo.: Roman Catholic Books, 1953.

Nickell, Joe. *Inquest on the Shroud of Turin—Latest Scientific Findings.* Buffalo, N.Y.: Prometheus Books, 1983.

Zugibe, Frederick T. "The Man of the Shroud Was Washed." *Sindon N.S.* Quad. No. 1, June 1989.

———. "Pierre Barbet Revisited." *Sindon N.S.* Quad. No. 8, December 1995.

CHAPTER 8: HOW TO KNOW IF YOU'RE DEAD

Ad Hoc Committee of the Harvard Medical School to Examine the Definition of Brain Death. "A Definition of Irreversible Coma." *Journal of the American Medical Association* 205 (6): 85–90 (5 August 1968).

Bondeson, Jan. *Buried Alive.* New York: W. W. Norton & Company, 2001.

Brunzel, B., A. Schmidl-Mohl, and G. Wollenek. "Does Changing the Heart Mean Changing Personality? A Retrospective Inquiry on 47 Heart Transplant Patients." *Quality of Life Research* 1:251–56 (1992).

Clarke, Augustus P. "Hypothesis Concerning Soul Substance." Letter to the Editor, *American Medicine* II (5):275–76 (May 1907).

Edison, Thomas A. *The Diary and Sundry Observations of Thomas Alva Edison.* Edited by Dagobert D. Runes. Westport, Conn.: Greenwood Press, 1968.

Evans, Wainwright. "Scientists Research Machine to Contact the Dead." *Fate,* April 1963, 38–43.

French, R. K. *Robert Whytt, The Soul, and Medicine.* London: Wellcome Institute of the History of Medicine, 1969.

Hippocrates. *Places in Man.* Edited, translated, and with commentary by Elizabeth M. Craik. Oxford: Clarendon Press, 1998.

Kraft, I. A. "Psychiatric Complications of Cardiac Transplantation." *Seminars in Psychiatry* 3:89–97 (1971).

Macdougall, Duncan. "Hypothesis Concerning Soul Substance Together with Experimental Evidence of the Existence of Such Substance." *American Medicine* II (4):240–43 (April 1907).

———. "Hypothesis Concerning Soul Substance." Letter to the Editor, *American Medicine* II (7): 395–97 (July 1907).

Nutton, Vivian. "The Anatomy of the Soul in Early Renaissance Medicine."

In *The Human Embryo: Aristotle and the Arabic and European Traditions*. Exeter, Devon: University of Exeter Press, 1990.

Pearsall, Paul. *The Heart's Code: Tapping the Wisdom and Power of Our Heart Energy*. New York: Broadway Books, 1998.

Rausch, J. B., and K. K. Kneen. "Accepting the Gift of Life: Heart Transplant Recipients' Post-Operative Adaptive Tasks." *Social Work in Health Care* 14 (1):47–59 (1989).

Roach, Mary. "My Quest for Qi." *Health*. March 1997, 100–104.

Tabler, James B., and Robert L. Frierson. "Sexual Concerns after Heart Transplantation." *Journal of Heart Transplantation* 9 (4):397–402 (July/August 1990).

Whytt, Robert. *The Works of Robert Whytt, M.D., Late Physician to His Majesty*. Edinburgh: 1751.

Youngner, Stuart J., et al. "Psychosocial and Ethical Implications of Organ Retrieval." *New England Journal of Medicine* 313 (5):321–23 (1 August 1985).

CHAPTER 9: JUST A HEAD

Beaurieux. *Archives d'Anthropologie Criminelle*. T. xx, 1905.

Demikhov, V. P. *Experimental Transplantation of Vital Organs*. New York: Consultants Bureau, 1962.

Fallaci, Oriana. "The Dead Body and the Living Brain." *Look*, 28 November 1967.

Guthrie, Charles Claude. *Blood Vessel Surgery and Its Applications*. Reprint, with a biographical note on Dr. Guthrie by Samuel P. Harbison and Bernard Fisher. Pittsburgh: University of Pittsburgh Press, 1959.

Hayem, G., and G. Barrier. "Effets de l'anémie totale de l'encephale et de ses diverses parties, étudies à l'aide la décapitation suivie des tranfusions de sang." *Archives de physiologie normale et pathologique*, 1887 Series 3, Volume X. Landmarks II. Microfiche.

Kershaw, Alister. *A History of the Guillotine*. London: John Calder, 1958.

Laborde, J.-V. "L'excitabilité cérébrale après décapitation: nouvelle expériences sur deux suppliciés: Gagny et Heurtevent." *Revue Scientifique*, 28 November 1885, 673–77.

———. "L'excitabilité cérébrale après décapitation: nouvelle recherches physiologiques sur un supplicié (Gamahut)." *Revue Scientifique*, July 1885, 107–12.

———. "Recherches expérimentales sur la tête et le corps d'un supplicié (Campi)." *Revue Scientifique*, 21 June 1884, 777–86.

Soubiran, André. *The Good Doctor Guillotin and His Strange Device.* Translated by Malcolm MacCraw. London: Souvenir Press, 1964.

White, Robert J., et al. "Cephalic Exchange Transplantation in the Monkey." *Surgery* 70 (1):135–39.

———, et al. "The Isolation and Transplantation of the Brain: An Historical Perspective Emphasizing the Surgical Solutions to the Design of These Classical Models." *Neurological Research* 18:194–203 (June 1996).

CHAPTER 10: EAT ME

Bernstein, Adam M., Harry P. Koo, and David A. Bloom. "Beyond the Trendelenburg Position: Friedrich Trendelenburg's Life and Surgical Contributions." *Surgery* 126 (1):78–82.

Chong, Key Ray. *Cannibalism in China.* Wakefield, N.H.: Longwood Academic, 1990.

Garn, Stanley M., and Walter D. Block. "The Limited Nutritional Value of Cannibalism." *American Anthropologist* 72:106.

Harris, Marvin. *Good to Eat.* New York: Simon & Schuster, 1985.

Kevorkian, Jack. "Transfusion of Postmortem Human Blood." *American Journal of Clinical Pathology* 35 (5):413–19 (May 1961).

Le Fèvre, Nicolas. *A Compleat Body of Chymistry.* Translation of *Traicté de la chymie,* 1664. New York: Readex Microprint, 1981. Landmarks II series. Micro-opaque.

Lemery, Nicholas. *A Course of Chymistry.* 4th edition, translated from the 11th edition in the French. London: A. Bell, 1720.

Peters, Hermann. *Pictorial History of Ancient Pharmacy.* Translated and revised by William Netter. Chicago: G. P. Engelhard, 1889.

Petrov, B. A. "Transfusions of Cadaver Blood." *Surgery* 46 (4):651–55 (October 1959).

Pomet, Pierre. *A Compleat History of Druggs.* Volume 2, Book 1: Of Animals. Third edition. London, 1737.

Read, Bernard E. *Chinese Materia Medica: Animal Drugs.* From the *Pen Ts'ao Kang Mu* by Li Shih-chen, A.D. 1597. Taipei: Southern Materials Center, 1976.

Reuters. "Court Releases Crematorium Cannibals." Oddly Enough section. 6 May 2002.

————. "Diners Loved Human-Flesh Dumplings." *Arizona Republic,* 30 March 1991.

Rivera, Diego. *My Art, My Life: An Autobiography.* Reprint. Mineola, N.Y.: Dover, 1991.

Roach, Mary. "Don't Wok the Dog." *California,* January 1990, 18–22.

————. "Why Doesn't Anyone Have Dropsy Anymore?" Salon.com, 2 July 1999.

Sharma, Yojana, and Graham Hutchings. "Chinese Trade in Human Foetuses for Consumption Is Uncovered." *Daily Telegraph* (London), 13 April 1995.

Tannahil, Reay. *Flesh and Blood.* Briarcliff Manor, N.Y.: Stein & Day, 1975.

Thompson, C. J. S. *The Mystery and Art of the Apothecary.* Philadelphia: J.B. Lippincott, 1929.

Walen, Stanley, and Roy Wagner. "Comment on 'The Limited Nutritional Value of Cannibalism.'" *American Anthropologist* 73:269–70 (1971).

Wootton, A. C. *Chronicles of Pharmacy.* London: Macmillan, 1910.

Zheng, I. *Scarlet Memorial: Tales of Cannibalism in Modern China.* Translated by T. P. Sym. Boulder, Colo.: Westview Press, 1996.

CHAPTER 11: OUT OF THE FIRE, INTO THE COMPOST BIN

Mills, Allan. "Mercury and Crematorium Chimneys." *Nature* 346:615 (16 August 1990).

Mount Auburn (Massachusetts) Cemetery Scrapbook I, page 5. "Disposing of Corpses: Improvements Suggested on Burial and Cremation." Article from unnamed newspaper, 18 April 1888.

Prothero, Stephen. *Purified by Fire: A History of Cremation in America.* Berkeley and Los Angeles: University of California Press, 2001.

CHAPTER 12: REMAINS OF THE AUTHOR

O'Rorke, Imogen. "Skinless Wonders: An Exhibition of Flayed Corpses Has Been Greeted with Popular Acclaim and Moral Indignation." *The Observer* (London), 20 May 2001.

United Press International. "Boston Med Schools Fear Skeleton Pinch: Plastic Facsimiles Are Just Passable." *Chicago Tribune,* 15 June 1986. Final Edition.

How to Donate Your Body to Science

Donating your body to science is different from donating organs for transplant. Being an organ donor involves marking your driver's license or adding yourself to the National Donate Life Registry (www.registerme.org). Then let your family know that you want your organs donated in the event you wind up terminally ill on life-support or braindead on a respirator (which keeps the organs viable for a few days pending the transplant).

To donate your body to a medical school or university for research, you'll need to contact the facility. Try a Web search for "body donation program" and the name of your state. Contact the institution you'd like to donate yourself to and request a body donation form and information packet. Be sure to talk to your family about your plans. For one thing, they'll need to call the university to come get you when the time comes.

You can't specify what you're used for; you go where there's a need. The majority of donated bodies wind up in the anatomy department. Almost none end up in the English department. Medical conferences where surgeons practice new techniques are another common venue. If there's something you'd rather *not* be used for, make this clear on your donor form or in an attached note. Have fun!

STIFF

Mary Roach

STIFF

Mary Roach

DISCUSSION QUESTIONS

1. In her introduction to *Stiff*, Mary Roach remarks that "death makes us helplessly polite" (p.13). Why is it that we're compelled to use polite language when discussing death? Why are we often afraid to discuss it in the way Roach has done here?

2. Roach discovered that students in anatomy classes tend not to enjoy touching and smelling cadavers, even though they relish the opportunity to study them. Does this surprise you? Why might someone want to work with cadavers?

3. Could one remain more psychologically and emotionally balanced in their dealings with cadavers by humanizing them, as Roach frequently does, or by objectifying them? Explain.

4. Roach describes the smell of a decomposing human: "It is dense and cloying, sweet but not flower-sweet. Halfway between rotting fruit and rotting meat" (p. 70). But modern embalming methods allow us to present odorless, good-looking corpses at funerals. Has modern mortuary science made death more aesthetically pleasing?

5. Dennis Shanahan, who investigated the grisly human wreckage of downed TWA Flight 800, told Roach that the hardest thing about examining Flight 800 was that most of the bodies were relatively whole. He said, "Intactness bothers me much more than the lack of it" (p.116). Why might he feel this way? Do you agree or disagree?

6. Many research studies that make use of cadavers raise questions about maintaining the dignity of the deceased. For example, a ballistics study might involve decapitating a cadaver or shooting one

in the face—all for the sake of gathering data to ensure that innocent civilians who are hit in the face with nonlethal bullets won't suffer disfiguring fractures. Do you think that the humanitarian benefits of experimenting on cadavers can outweigh any potential breach of respect for the dead? Why or why not?

7. The heart, cut from the chest, can keep beating on its own for as long as a minute or two. This, Roach says, reflects centuries of confusion over how exactly to define death. Have modern scientific experiments on cadavers helped us to pinpoint the precise moment when life ceases to exist and all that's left is a corpse? Explain.

8. Roach says, "On a rational level, most people are comfortable with the concept of brain death and organ donation. But on an emotional level, they may have a harder time accepting it" (p. 188). Some organ recipients even worry that they will take on certain characteristics of their donors. What might this say about how we link the physical human body to the human soul?

9. In Chapter 10, Roach takes us on a grand tour of cannibalism across cultures. She's compelled by the idea that economics accounts for why people throughout history have never dined regularly on each other. Humans, she says, turn out to be lousy livestock, because you have to give them more food to feed them than you'd gain in the end by eating them. How do you react to this idea?

10. In Chapter 11, Roach journeys to an island in Sweden, where a forty-seven-year-old biologist-entrepreneur has made a business of producing compost from cadavers. This business has major corporate backing and an international patent, and mortuary professionals in many countries, including the United States, are interested in representing the new technology. Do you think that the "human compost movement" could gain traction where you live?

11. Roach concludes that "it makes little sense to try to control what happens to your remains when you are no longer around to reap the joys or benefits of that control" (p.290). Do you agree with her?